"You see things and you say 'Why?'.
But a designer dreams things that
never were and says 'Why not!'."

George Bernard Shaw, modified by Dick Powell

RotoVision

A RotoVision Book

Published and distributed by RotoVision SA
Route Suisse 9
CH-1295 Mies
Switzerland

RotoVision SA
Sales and Editorial Office
Sheridan House, 114 Western Road
Hove BN3 1DD, UK

Tel: +44 (0)1273 72 72 68
Fax: +44 (0)1273 72 72 69
www.rotovision.com

10 9 8 7 6 5 4 3 2 1

ISBN: 978-2-88893-148-5

Art Director: Tony Seddon
Design: Jane Waterhouse

Reprographics in Singapore by ProVision Pte.
Tel: +65 6334 7720
Fax: +65 6334 7721

Printed in China by Midas Printing International Ltd.

Issues

Anatomy

Portfolios

Etcetera

What is product design?

As it is such a vast subject area, this book showcases a personal selection of products that offer a useful insight into design considerations and processes. Maybe we should ask, what are the main facets that make up a product, its life cycle, and market? A product can be physical or metaphysical. This book attempts to uncover the generic ideas behind the design of products while identifying physical and intellectual issues pertaining to product design.

Product design is an ambiguous term that blurs the boundaries between specialist fields of lighting, furniture, graphic, fashion, and industrial design. Perceptual and physical boundaries constantly erode as global communication improves. Boundaries are areas of activity where dichotomies and tensions spark communication, confrontation, and creativity. This is evident in the multi-disciplinary objectives of many designers' portfolios, and in the cooperation between designers from different stylistic and specialized fields in creating products. However, there still are (and always will be) highly specialized areas of product design that are niche oriented and skill specific.

Are designers the new futurologists? No, but designers need to be in line with and to meet people's needs. They should identify clearly the relationships between the past, present, and future, and the potential effects of political, social, and emotional influences within their environment. Technological developments in logistics and information systems bring global neighbors closer to each other and enable them to communicate and exchange cultural values through many vehicles of expression.

Design is a form of expression and products are a currency for exchange. Products as a currency generate revenue. Perceptions of design within business have evolved and are now key in the corporate arena as companies find that investment in design gives returns on investment. A carefully designed and marketed product can bring iconic status to a company or designer. It can also offer a unique stance in a highly competitive world.

Product design is a generic term for the creation of an object that originates from design ideas—in the form of drawings, sketches, prototypes, or models—through a process of design that can extend into the object's production, logistics, and marketing. Products are designed with particular considerations valued by the designer, client, or end user that are then communicated through the products' purchase and use. Product life cycles shorten as fashion and technological improvements affect our product selection criteria. How can we create in more innovative and challenging ways to embrace new social trends?

Products help us to interpret our positioning in the world. This can lead to a skew toward a social context in product design. Now that we have a plethora of product choices, how can designers improve an existing product or offer something new? What is a new product? As we embrace internationalization and as product development cycles quicken, designers need to take a fresh approach to the question "what is product design?" Is it a designer's responsibility to be designing for our future and to find new and effective ways of producing what people want?

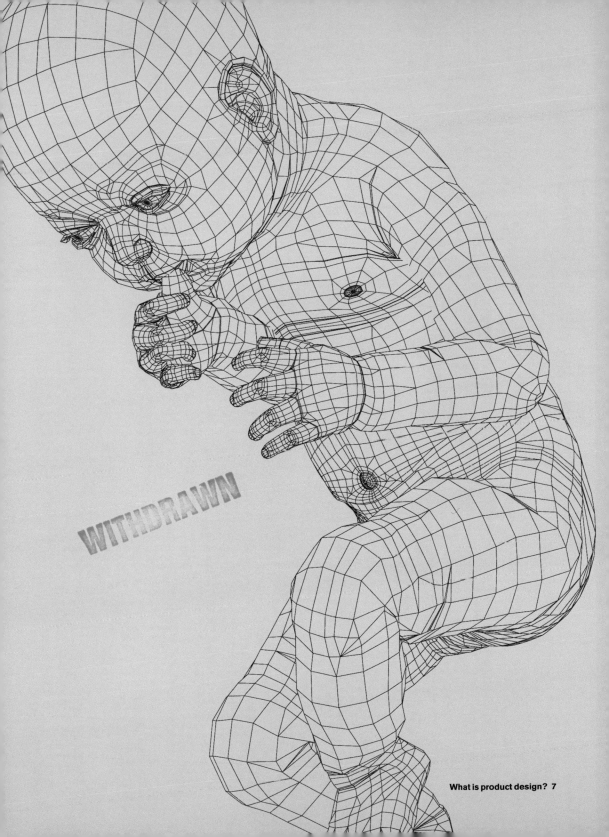

WITHDRAWN

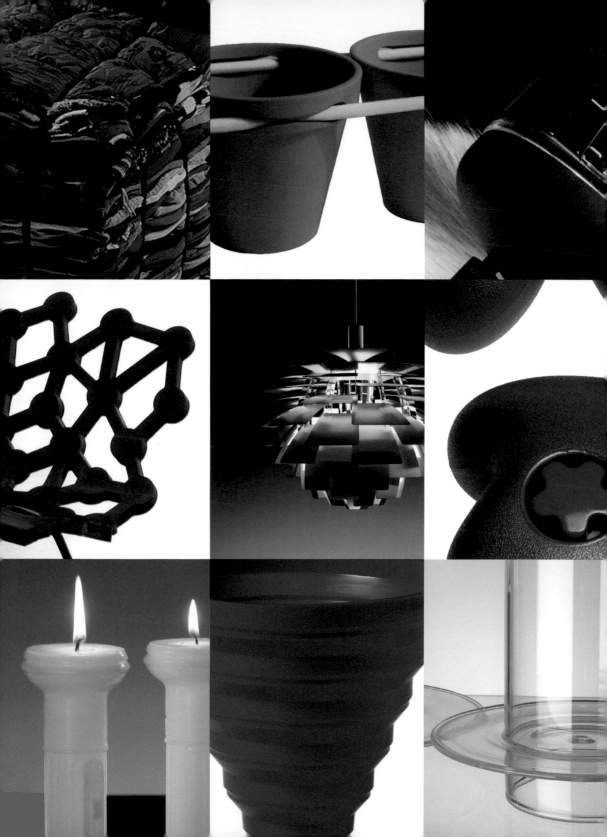

The design story: how product design evolved

Design is an ancient and historic activity that dates back to early civilizations. The first products were hand crafted and served a primary function for use within a geospecific community. Personalization through arts-and-crafts skills brought greater desirability and the potential of exchange for revenue. Consider the Japanese lantern from the Edo Dynasty, in which decoration takes the primary function of illumination to an esthetic dimension with hand-painted detail. An object becomes a desirable product when it gives the appearance of having improved functionality; additional cultural meanings or additional processes in design or production create an increased demand for a product. Demand requires improved production methods to enable adequate supply.

As indicated on our timeline chart on page 14, certain periods show how the history of design is directly related to the history of worldwide economic development. This timeline shows the well-known design movements.

Industrial design appeared during the 1920s and 1930s as a result of the Industrial Revolution, which in turn created new divisions of labor. The changing work environment brought about by early methods of mechanization and improved product output induced a greater degree of specialization within the workplace. The idea of a professional designer emerged.

Art Nouveau in France, Jugendstil in Germany, Modern Style in the UK, and Modernism in Spain brought design movements to the forefront of stylish products relevant to each benchmark in time. The Bauhaus school, active from 1919–1933, introduced new ways of teaching industrial design—the application of equal measures of art and technology. The process of design questioned the design and technological boundaries in hands-on workshops within the learning environment. Some of these principles are still applied today within higher education. Toward the end of WWII, a new style appeared, based on architecture and consumer goods being functional and affordable due to mass production. There was also a unification between arts, crafts, and technology that resulted in high-end functional goods entering the marketplace. Then the term product design related to the principle of product—the generation of multiples of the same item. With this ease of production, there was an addition to the design process that applied styling and planned obsolescence. The main themes in these stylistic periods and their related products offer some insight into the theories then posed. The consumer boom started in the 1950s when there was a move toward functionalism with practical and economical design. Architectural designers created interior products to suit their buildings and synthesize a certain look that complemented architectural structures. The 1960s brought a popular counterculture to the 1950s ethos, using humor and the fusion of diverse materials that were inappropriate for the 1950s generation. The emphasis was on discovering new techniques and forms of expression, and also on taming new materials and processes.

During the late 1970s and 1980s, an international bonding of like-minded consumer groups created new challenges for designers—the eventual global concept emerged in the 1990s. Product and industrial designers explored design within different disciplines and key designers forged their own career paths. Industrial design and product design were synonymous design classifications that resulted from the progressive development of designers as artists mirroring social considerations and desires, while applying technical expertise to production methods and materials.

So how has design evolved today? There are many terms that give us some insight into today's design positioning within the multicultural world. Buzzwords that represent the here and now become out of date as quickly as they are coined. The quest for individuality and mass appeal will always be the designer's focus, with an everlasting note to self—keep an eye on the present, understand the past, and look toward the future. There are no set patterns, words, or phrases that best explain this reality.

Above: Swiss Army Knife
Designed by Victorinox.
Originally produced as part
of the Swiss Army kit in 1891,
this knife has since been
adopted as a useful, everyday
object in mainstream society.

Left: Cylindaline tea pot
Designed by Arne Jacobsen.
Part of the 1967 Cylindaline
collection of satin-polished
stainless steel products,
which is still in production.

Balloon lamps
Designed by Kyouei.
Traditional paper lanterns
are still used today, but
this modern version uses
a balloon, LED, battery,
and attachment device.

Left: Bubble lamps
Designed by George Nelson. Nelson designed a range of bubble lamps, including wall sconces, floor lamps, and pendants, with the first produced in 1947. The soft light that shines through emphasizes their simple, sculptural appeal.

Below: Arco lamp
Designed by Achille and Pier Giacomo Castiglioni. This lamp has reached iconic status in the design industry as one of the best-known floor lamps. A directional light with a white Carrara marble base; satin-finish stainless-steel telescopic stem; and pressed, polished, zapon-varnished aluminium adjustable reflector. Sold globally and still in production. A limited edition of black marble, signed and numbered pieces were produced to commemorate the passing of Achille.

ICONIC PRODUCTS

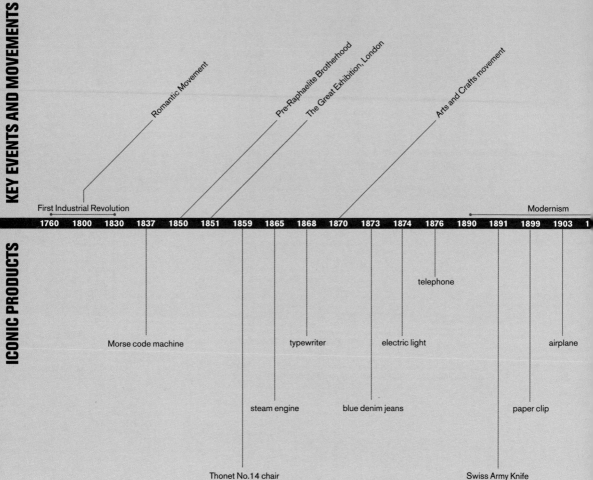

Romantic Movement

Pre-Raphaelite Brotherhood

The Great Exhibition, London

Arts and Crafts movement

First Industrial Revolution

Modernism

| 1760 | 1800 | 1830 | 1837 | 1850 | 1851 | 1859 | 1865 | 1868 | 1870 | 1873 | 1874 | 1876 | 1890 | 1891 | 1899 | 1903 | 1 |

telephone

Morse code machine

typewriter

electric light

airplane

steam engine

blue denim jeans

paper clip

Thonet No.14 chair

Swiss Army Knife

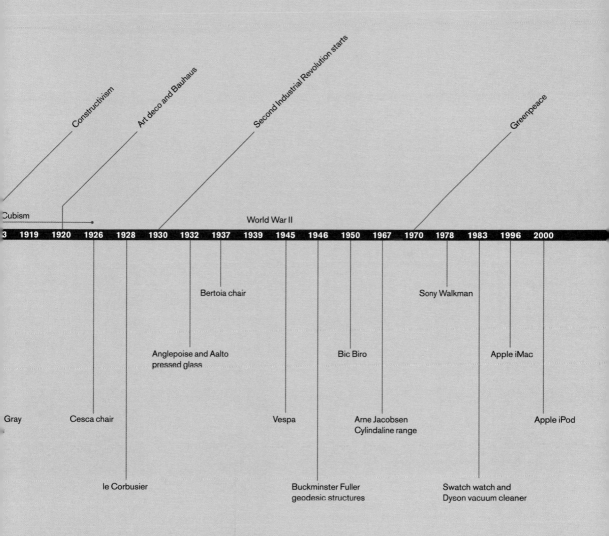

Constructivism

Art deco and Bauhaus

Second Industrial Revolution starts

Greenpeace

Cubism

World War II

3 1919 1920 1926 1928 1930 1932 1937 1939 1945 1946 1950 1967 1970 1978 1983 1996 2000

Bertoia chair

Sony Walkman

Anglepoise and Aalto
pressed glass

Bic Biro

Apple iMac

Gray

Cesca chair

Vespa

Arne Jacobsen
Cylindaline range

Apple iPod

le Corbusier

Buckminster Fuller
geodesic structures

Swatch watch and
Dyson vacuum cleaner

New design: an era of hybrid products

A hybrid product is formed when two or more objects are merged to create something new, but that still has visible references to a past life. The act of hybridizing is a provocative way of questioning the relevance of an object, and signifies an element of progressive thought by which we challenge what is the basic truth behind the reason for the product itself. Certain qualities from each original object can be enhanced or subdued to give the desired effect. Consider the use of a wine glass as a doorbell. What do you say when ringing the doorbell? Good health my friends and a long life? A wine glass makes a wonderful shrill sound when toasting friends. With a wine glass doorbell we embrace the visual and also the acoustic elements of a product; sight is posited with sound.

There have been many products developed through this approach to design, and through them we experience new ways of conveying functionality, esthetics, and meanings. Designs reveal their character through artistic expression. This different approach to design shows how influential the actual design process can be, as opposed to the constant need for completely new objects. Why not apply methods of reappropriation or recontextualization? These principles are simply a way of changing how we view, use, and interact with objects.

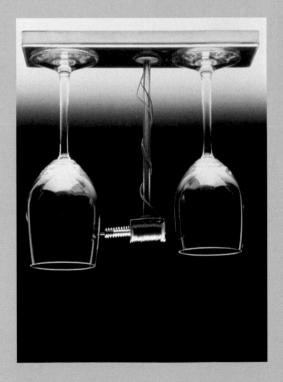

Above: Fu Ring
Designed by Mari Isopahkala.
Silver and trimmed reindeer
fur. This design was part of the
project "30 For Sale" at the
University of Art and Design,
Helsinki. The usual application
of reindeer fur as a by-product
is for flooring or slippers, but
here this material is still in a
natural state and has its own
particular form as a beautiful
piece of jewelry.

Left: Bulb
Designed by Paul Cocksedge.
Placing a cut flower into the
vase of water completes the
electrical circuit and produces
light. The water helps diffuse
the light while also nourishing
the flower. The shape and
physical ingenuity of this
design references a laboratory
setting where the designer
becomes alchemist.

Right: Sella stool
Designed by Achille and Pier
Giacomo Castiglioni in 1957.
The Sella (Italian for saddle)
stool was designed for a
posture somewhere between
sitting and pacing nervously,
to accommodate Achille's
desire for a comfortable seat
to make phone calls from: he
liked "to move around" and
"to sit, but not completely."

Functionality is for fun

People are becoming more sophisticated in their expectations of and needs for products. The idea of entropy is relevant; it is natural for life and all things to become more chaotic with time. Designers engineer and infuse an object with perceptual and physical values that reach beyond pure functionality.

Function is often posited with form, yet such a purist approach to design seems to be less popular than the idea of layering other value-generating product offerings that go beyond form and function. An emphasis on other aspects of the object can induce a novel product through greater attention to intangible qualities. What is the nature of function? What do we mean by this term? As products utilize multifunctionality and modular interaction with end users making final configurations, functionality is now hidden and less complex. The marriage of certain characteristics makes a product stand out and even suggests a less recognized use. Ambiguity and challenging the traditional norms of form and function are the mainstays for today. Is functionality for fun? We still and always will need greater functionality within certain product categories.

There is something calming about a product with great visual and perceptual honesty within the principles of form, as it bestows the observer with clarity about its function and use. Pure functionality in a sleek and fluid form can be seen with the nature-bound sport of surfing. A surfboard is functionality for fun. Fluid dynamics and ergonomics using lightweight materials result in water acrobatics and fun. The surfboard is an extension of the self, as board and man need to find affinity to ride waves.

Once again the question of unpredictability arises; the idea of function can mean many things and can offer an area for redesign in a conceptual format. Industrial Facility uses the traditional sunglasses shape for a hair band; Shin and Tomoko Azumi apply visual play through cartoon characters with the smile on their Snowman salt- and pepper shakers.

AUTOBAND
Design: Martí Guixé
Editor: Galeria H O Barcelona

Left: Autoband
Designed by Martí Guixé.
With this application the value
of sticky tape is changed from
pure functionality to an
interactive game.

Above: Puppy
Designed by Eero Aarnio
(photography Carlo Lavatori).
Abstract dog in rotational,
molded polyethylene. In this
product functionality is
hidden; this design is not
just an ornament or toy, but
also a seat. Through the
simplification of shape and
rationalization of facial
characteristics, the dog
is depersonalized, yet has
a friendly appearance.

Timesphere
Designed by Gideon Dagan.
A seemingly gravity defying
ball travels round the dial to
mark the time.

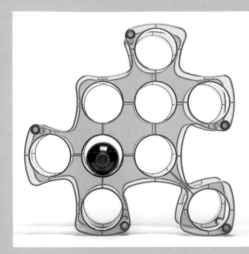

Puzzle Wine Rack

Designed by Gideon Dagan. Injection-molded ABS (Acrylonitrile Butadiene Styrene) polymer in a range of colors. The units can be snapped together to create the desired wine-rack size. One unit offers two placement positions and can hold up to nine bottles.

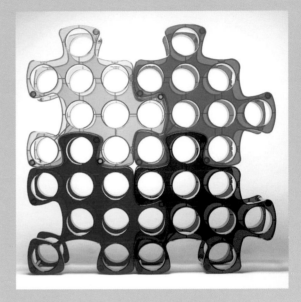

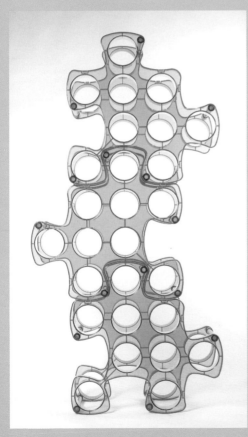

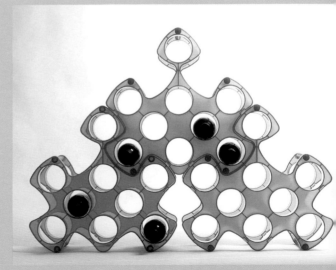

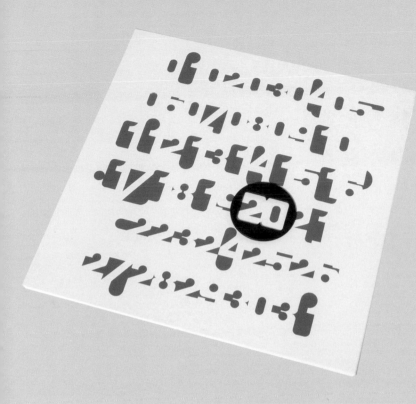

Opposite: Gemini candleholder
Designed by Peter Karpf. Interactivity adds value as the end user can define this repeating graphic pattern to form a functional candleholder.

Left: Imbroglio calendar
Designed by Jean Pierre Vitrac. The joy of this calendar is the "hidden" nature of the numbers. It is only with the positioning of the plastic cursor that each number becomes clear. (Imbroglio means hidden or muddled in French.)

Below: Solitaire dish
Designed by Barnaby Barford and Andre Klauser. Is it a game of Solitaire or a dish for olives? Functionality is for fun.

Emotional ergonomics and hidden functionality

Today, technology develops constantly, regardless of any key sociopolitical benchmarks. Buckminster Fuller noted that inventions seemed to melt down roughly every 25 years, after which the original materials recirculate in novel and more effective applications.

Technical innovation is one of the long-known motivators and facilitators within the industry. The main areas of change and consideration have been in the development of new materials, components, applications, and information resources. Developments in materials offer a multitude of choices for application—how do you choose a specific material and what properties of the material are paramount? How to select the right material will relate directly to prioritizing set facets of the products as well as the softer, less tangible qualities of end-user perception.

Materials take on the definition of the product's skin; they conceal its internal workings and are the human interface. At present there is a move toward this design interface with a sensory focus on material tactility for more high-tech products.

Emotional ergonomics juxtaposes a highlighting of the sensory perception of touch with a paring down of the perception of sight. Technological advancements in the material processes of shaping and production enable a designer to convey the smooth surface of a sleek skin which feeds the brain with sensory emotions. The rationalization of material and functional complexity creates a different wow factor.

Technological developments in product components have led to their size reduction. Chip, microprocessor, and nanotechnologies make the function-related components within a product less noticeable in the final design, allowing the designer more freedom in the product's configuration. Empirical expressions of the functional aspects offer quick response and easy-to-use functionality. Sensory stimuli engineered into products influence conscious and unconscious product selection.

Right: iPod
Designed by Apple. Material components have a visual ghost-like quality yet at the same time stimulate to touch and feed the human senses. Small audio and tele-communications products adopt an emotional context as human senses are engineered into the products. Functionality is hidden or has a minimalist interface for product responsiveness.

Left: Isokon Stool/ Round Tray
Manufactured by Isokon Plus, the Isokon Stool/Round Tray was originally designed in 1933. This versatile design can be used as a stool, tray, or occasional table.

Right: Miroir
Designed by Christopher Pillet and made exclusively for Galerie Kreo. Black, anodized aluminum and mirrored glass are used to create an ambiguous form that is both a sculpture and a hand-held mirror.

Coffee Maker

Designed by Jasper Morrision for Rowenta. Tucked away and out of sight you find all the equipment needed to make coffee—there's no need for a spoon or measure and the filters fit perfectly within this compact design. There's greater interactivity and a degree of surprise as you locate the flip-top store. Understated form and color means this usually rather cumbersome product type sits quietly on the side. You notice it for its honesty and secrecy.

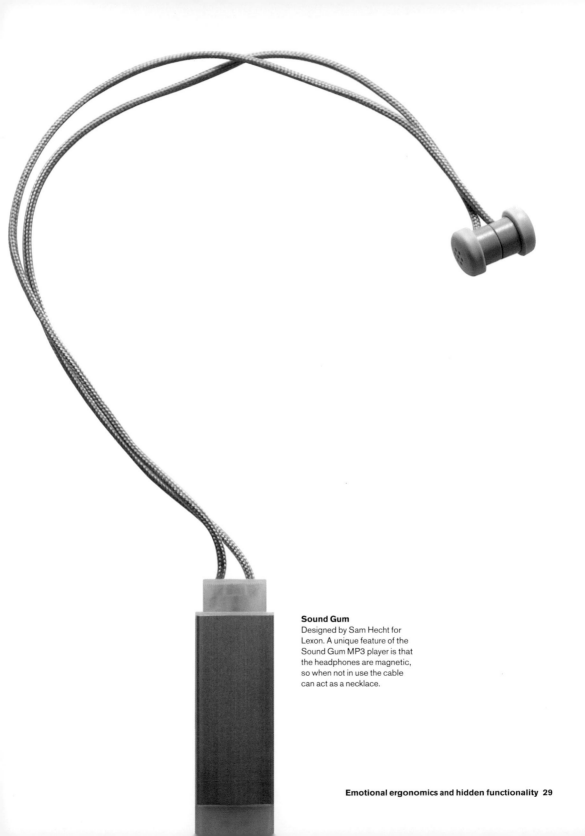

Sound Gum
Designed by Sam Hecht for
Lexon. A unique feature of the
Sound Gum MP3 player is that
the headphones are magnetic,
so when not in use the cable
can act as a necklace.

Informed design and communication

Technological developments influencing the flow of information through the different facets of supply enable the designer to get closer to the end user. In the trickle-down theory of fashion, the flow of information usually moves only in the direction of the end user; now there is access to detailed information about all departments within the design process and, more importantly, on the customer. A designer should be provided with as much relevant information as possible about their customer's product requirements in order to create the best design solution. Here the customer can be a manufacturer, a design company, or the end user. Research helps to define and refine the prioritizing of the customer's perspective on key issues of specific product facets and can be supplied easily within today's information age.

Information also needs to flow freely between the different departments of design, production, marketing, accounting, and customer liaison. A more holistic approach to this flow of information helps all departments become more efficient, and can help to pinpoint potential areas for further design research and development. Product diversification for an existing design can be achieved. Sometimes whole product areas are found to have large voids in which there is not enough product choice. Those who can find new ways of becoming closer to and identifying the needs and demands of their customers will benefit the most. Developing better relations with all departments and end users can offer greater understanding of product needs. Omlet's Eglu was launched after a long period of research and prototyping. Ten prototypes were made to ensure that the look and structure met certain practical needs and criteria. These criteria were identified through building communication between the company and customers through rigorous testing, and even filming inside an Eglu to understand issues such as animal welfare. Feedback on the overall look and functionality of the design was gathered by interviewing those who bought the early models. This informed design approach affected the selection of materials and colors, and the composition of the Eglu. One color was replaced with a more natural, less toylike version due to customer feedback.

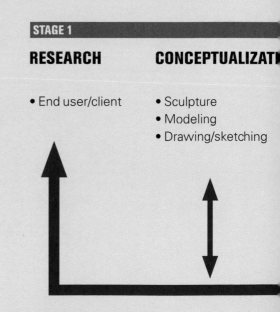

STAGE 1

RESEARCH

CONCEPTUALIZATI

- End user/client

- Sculpture
- Modeling
- Drawing/sketching

Flo Creator Model
Created by Flo design
consultancy. Information
resources are areas of
idea generation and
revenue formation.

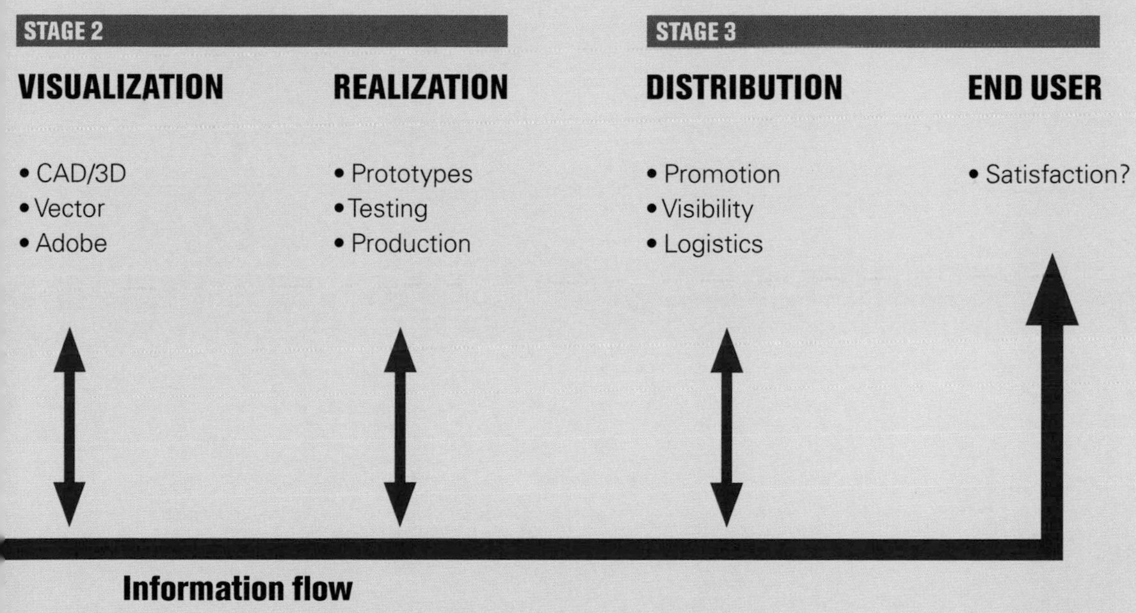

| STAGE 2 | | STAGE 3 | |

VISUALIZATION REALIZATION DISTRIBUTION END USER

- CAD/3D
- Vector
- Adobe

- Prototypes
- Testing
- Production

- Promotion
- Visibility
- Logistics

- Satisfaction?

Information flow

Left: Lux It ceramics
Designed by Alex Estadieu. Here, traditional ceramics, metal dishes, and takeout containers are deconstructed and reformed to take shape as an interpretation of our dining habits. Research into end-user habits provide inspiration for new products. Photo by Chikako Harada.

Above and right: Omlet's Eglu
Designed by James Tuthill, Johannes Paul, Simon Nichols, and William Windham. Rotational molding made the Eglu affordable and easy to clean, and also provided a solution to keeping the animals warm. The air between the double-layer skin helps maintain a constant temperature inside the Eglu.

Dual and multifunctionality

Dual or multifunctionality is demanded by urban life, where space is expensive and the principle of work is challenged by leisure time and flexible working hours. Added value can now mean additional functionality with perceived improvements in time- and space-saving properties. Static forms that once served a single purpose now provide dual or multifunctionality; perhaps a side table becomes a tray, a storage facility, or even a vase. The simple way the end user can interact with the object is not confusing, but beneficial as the product reinforces its rightful presence within a space. Our homes are now also places of work. The way we split time and the space in which we live reveals potential for product development in helping to tackle working from home.

High-tech products have been offering additional facets of use for some time—more hand-held devices than ever are being bought. Communication methods have improved through product development and cross-functional engineering to give greater diversity and allow quick and easy interaction with friends and family; most cellphones also have a camera and e-mail function.

Chameleon products either have a function that is out of sight when not in use, or change to move from one use to another, producing a more ambiguous effect. Still other products make a show of this added functionality. Why not have a product that offers several solutions to the way we live? Do you pay more for an extra function? Is it time-saving or just a gimmick? An open bag of coffee, sugar, flour, or anything needs a seal, but then why have a separate spoon? The Spoonclip, fashioned for dual use, works.

Until we understand the similarities and regularities that occur within our daily existence, many multifunctional products will miss the mark and just become the latest gimmick. Some items offer a real value and can reduce the need for other products as they perform well with more than one end use, or offer a long-term use.

Above: X-table
Designed by James Irvine for Swedese. Fold away for another day or use as a side table or tray. There is also the choice of using either the white or black side of the tray.

Right: Shelflife
Designed by Charles Trevelyan. This system integrates shelving of various proportions with a chair and workstation that are hidden within the structure. Photo by George Ong.

Below: CandleMaker
Designed by Designfenzider Studios. Placing these porcelain accessories in a wine glass converts a drinking vessel into a candle.

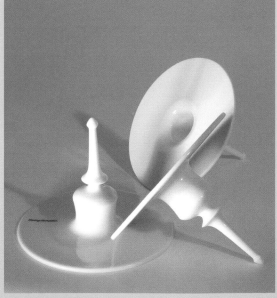

Left: Spoonclip
Designed by Jeremiah Tesolin. A spoon and clip are combined to provide a measure and seal for everyday household goods. Tesolin, a Helsinki-based Canadian designer, works in design and marketing for studios, large corporations, and his own practice. Photo by Chikako Harada.

Right: High Chair
Designed by Maartje Steenkamp for Droog Design. In our everyday lives eye contact is a necessity. A baby sits on a high chair in order to be at eye level with other diners. As the child grows, the chair needs to be reduced in height. Supplied with a saw and sanding paper, the height of this chair can be adapted in line with a child's growth.

Below: Epsistola letter scale
Designed by Teo Enlund for Simplicitas. This ingenious product is a combined letter scale, calibrated to postal rates, and letter opener. Photo by Joakim Bergström.

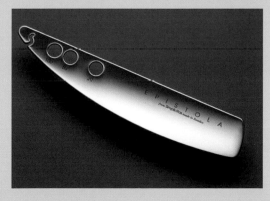

Decoration

Whereas technological products adopt modern simplicity, decorative products reveal a renaissance of ornate embellishment originating in nature or synthetic environments. There is a convergence between different specialist fields— both graphic designers and tattooists create wallpapers for your home. The conventional application of pattern repeats has, in the past, been the skill base of textile designers. Now specialists from many different fields apply their own style and interpretation of patterns to textiles and wallpapers.

Increased complexity in repeating pattern is possible using graphics packages. While this ornate graphic power has emerged since the late 1990s, there is also a hand-drawn, more crafted look that remembers past styles of ornamentation and production. Traditional production techniques give a handmade feel to products with their layering effect, adding depth and color. References to flora and fauna recur in design history and offer an insight into our need to be closer to nature using the symbols we know and love. Within contemporary design there is a definitive style that takes heed of these traditional symbols and narratives, which also reflects the greater levels of complexity and precision facilitated by computerization. Structures that repeat can be applied to different materials and end uses, creating a diverse range of lighting, wallpapers, tiles, posters, and textiles.

Another influence on pattern is the adoption of different methods of inducing imperfections into a design, while still applying computerized, mechanical, or handmade production processes.

A designer plays with perfection and "nonperfection" in creating a product, particularly in light of computerized assistance where there is a potential to be uniform. We also see the convergence of applications with part hand-drawn, part computerized layering to produce less rigid forms. Crossing boundaries between certain specialist areas of visual expression becomes more inventive with unique coalitions between artists and producers; tattooists and embroiderers create a wallpaper pattern that references both skill bases. The options are endless when you think of the different ways designers and producers can work together.

Left: Ant Damask wallpaper
Designed by Paul Simmons of Timorous Beasties. Produced for the International Contemporary Furniture Fair (ICFF) show in New York, this handprinted faux flock damask pattern is flooded with tiny ants.

Above: Walled Paper
Designed by Eric Barrett of Concrete Blond. By embossing traditional wallpaper patterns in concrete castings, Barrett provides a new esthetic for concrete with his Walled Paper designs.

Right: Knock-Knock wallpaper
Designed by Keith Stephenson of absolutezero°. Part of the Playtime series, each design is laser cut onto a roller and then printed via flexo machines.

Left and below:
Cushions and textiles
Designed by Selina Rose.
Fabrics are embellished
with embroidery, cut-out
foils, layered textiles, and
screenprint techniques.

Right: Camoflage Chair
Designed by Gudrun Lilja
Gunnlaugsdottir. Part of
a series of chairs designed
in keeping with their
environment. The series
includes a temporary
"sandcastle" chair at the
seaside; Chairwood, to
blend with the woods on
Fourfootmoor; and Frosen,
a chair encased in ice.

Below: The Ugglas
Designed by Eva Schildt.
Candle accessories,
fashioned from metal, are
designed to cast shadows
of distorted animal shapes.

**Above and right:
Gardener's Sofa**
Designed by Eva Schildt.
This bench provides a frame
on which plants can grow.
The design improves with age
as the vegetation becomes
more established.

Roses rug
Designed by Nanimarquina.
This 100% wool rug is formed
from repeating circular felt
pads. Photo by Albert Font.

Design morality

Within an exchange there is usually a winner and a loser, to a greater or lesser degree. Is not everything a dichotomy of tensions? As some companies and industries move production around the world to locate the greatest revenue returns for mass-produced designs and hand-crafted items, then a fair balance in sociopolitical and economic issues needs to be addressed.

Maybe the solution lies in the way we work, and by increasing our awareness of different approaches to problem solving within this issue of design morality.

Some designers and organizations work effectively with talented and specialized craftsmen to develop products that are created in mutually beneficial economies. Some traditional decorative methods are relaunched into product design with a fresh new perspective that creates interest with a degree of ambiguity and mystery. How did they make that? What is it made from? Where did it come from?

As we do not have endless resources, how does this impact on product designers and their designs? Some products are likely to be turned into waste long before their expected life span has been met. Maybe we can look to design classics that stand the test of time to question their longevity, or adopt new and more interesting approaches to the process of product design?

By balancing our need for new products and our concern for design morality, we can address ways to improve and create design methods. It is now that we should pay attention to design morality and even to the necessity of design.

There are means by which we can improve life spans, and not only by using ecological materials. We can consider: redesigning within the areas of waste management, reuse, restorative waste cycles in which one man's waste is another's fuel, ease of repair, durability, flexibility for choice, and add-on components with improved upgrades. We need to devise new methods of working and communicating beyond geo- and sociopolitical boundaries, to question material applications, and also take on the consequences of our imagination when design fails in some way.

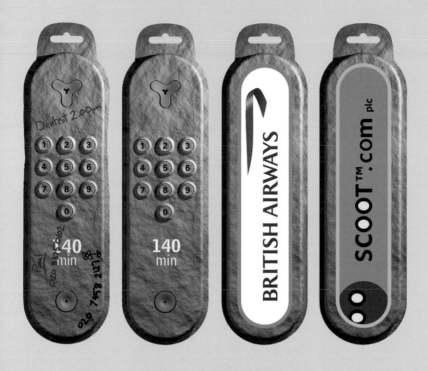

Left: Disposable mobile phone
Designed by Chris Christou at Youmeus Design Ltd. Recycled paper pulp—affordable, disposable, and recyclable—offers a pared-down functionality for a disposable mobile phone. You can even scribble a number on the phone itself. A visionary economical design.

Below: Garden Bench
Designed by Jurgen Bey. From grass cuttings to seat, the Garden Bench uses materials that are generally considered to be waste. The composter allows the creation of any length of bench, and the extruded compost is then fortified with resin to provide a rigid seat. Photo by Marsel Loermans.

Styrene light
Designed by Paul Cocksedge.
Polystyrene cups are reused,
hand crafted and shaped to
form an ambient pendant light.

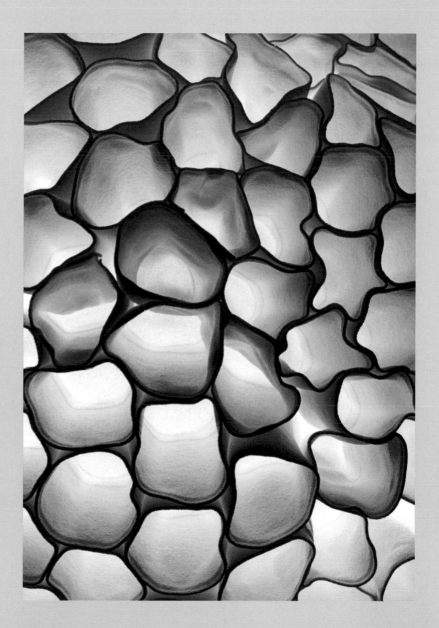

Paperbag
Designed by Jos van der Meulen. Meulen sews unused or misprinted billboard posters together to give them a new life as paperbags.

Birdhouse
Designed by Jeremiah
Tesolin. A birdhouse made
entirely out of bread provides
an edible, biodegradable
home; the product does not
have longevity, but it will not
become waste.

Formed sensibility

What is the common view? Is it how we look at objects as end users or as designers? Often you can use a product in a way other than that intended. How people use certain objects in their immediate surroundings offers international behavioral truths that underlie all our environments. The ability to see options for end use enables designers to find refreshing approaches to the generation of ideas and products. These sensibilities are in us all; we can observe minor differences in our everyday existence, in the way we hold a pencil, sit, or sleep.

Sculptural form, or any way that we can transform an idea into a beautiful shape, is an art that is encapsulated in different arenas of expression. Isami Noguchi used sculpture to find affinity with the forms he created by drawing and sketching shapes until they seemed right, or by modeling with computer technology. There are so many ways we can find a narrative for shape forming.

Patterns in nature help us to create structural shapes. Designers are influenced by mathematical principles present in nature, and subconsciously recognized. The Golden Mean, one such recognized principle, is used particularly in architecture. You recognize a shape when it seems right—this can be a visual consequence of natural mathematics.

The memory of other objects, symbolic recognition, helps us interpret use. Through this we know an object will perform a certain task or feel a certain way. Naoto Fukasawa linked a kitchen fan and a CD player through the pull-cord method of operating the device, which used this principle of recognition of another form and functionality. An invisible aspect of the design only becomes useful or interesting on interaction.

How can we develop sensibilities for other qualities in an object? The focus is often not on the object's characteristics, but on the subject. The behavior of the end user can add to the product's esthetic. This behavior can directly affect the physical appearance of the product; the Do Hit chair from Droog is designed to offer a personal touch through individual customization.

Ultimately, design is not just about the application of technology, form, and functionality; rather, it is about people and understanding behavior. Perhaps this is where we can locate the essence of designing the unexpected?

Milkbottle Lamp
Designed by Tejo Remy. Made from stainless steel, standard milk bottles, and bulbs, the inspiration for this piece sprang from a domestic setting. The bottles hang in a cluster of 12, three rows of four, exactly as they do in a Dutch milk crate. Photo by Marsel Loermans.

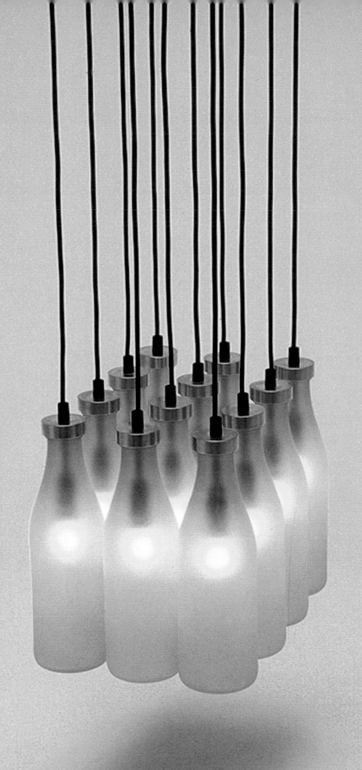

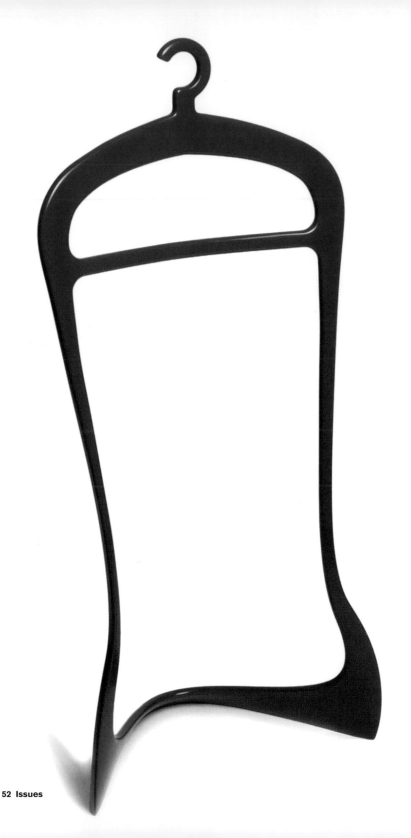

Left: Standing Hanger
Designed by Magnus Long.
A simple and elegant
fiberglass form that provides
an alternative to throwing your
coat over the back of a chair,
and allows you to stand damp
clothes in front of the radiator.
Photo by Mark Whitfield.

Top right: Do Hit
Designed by Marijn van der
Poll and produced by Droog
Design. With Do Hit the end
user can create their own
masterpiece: they refine the
final shape of the chair by
hitting it with a mallet.
Photo by Droog Design.

Bottom right:
Mary P Coat Rack
Designed by Eva Schildt.
Schildt designed this coat
rack with the image of water
in mind, and the chime of
the hooks is reminiscent
of a stream.

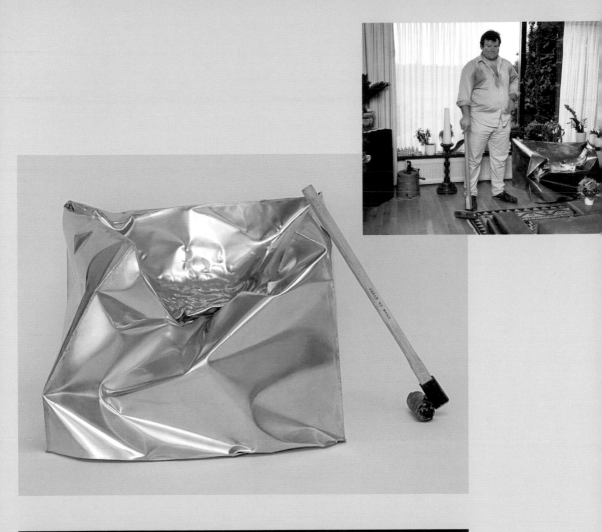

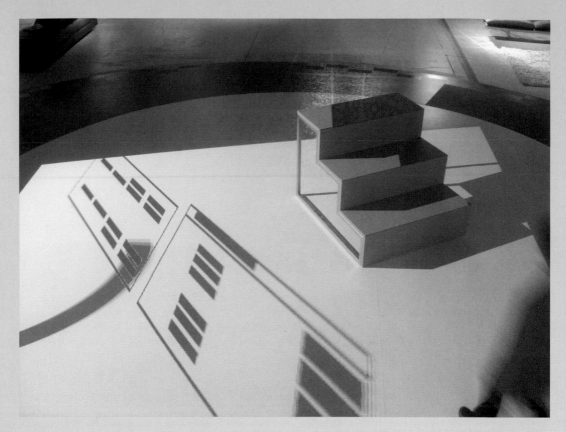

Above: Steps
Designed by Martí Guixé.
Often we sit on the stairs while
chatting with friends on the
telephone or at a party, hence
Guixé's staircase seat.

Right: Waterproof vase
Designed by Maxim Velcovsky
of Qubus. This porcelain vase
finds a new application for the
form of galoshes.

A new meaning: design as a statement

Postmodern fashion was concerned with the adoption of styles from the past and the reappropriation of clothing to give new meaning. New meanings can be physical manifestations or a layering of hidden messages. This concept seems to have been embraced within product design during the 21st Century. The clothes we wear, the products we buy, and the way we furnish our homes are expressive extensions of our selves. Some designers have taken a political stance and shout out to show the futility of the world in which we live. Design takes on the realms of art where a political, social, or cultural point is made through a product. The product becomes a communicator of these tensions. The application of gold on the Bedside Gun lamp designed by Starck offers an insight into the symbolic representation between the collusion of money and war.

The literal or indirect meaning or context of an image can conjure up issues and questions regarding changes within our society. Sometimes a statement within a visual image has a stronger, more lasting impact than words alone. On the other hand, a designer may look at an object and alter the overall perception of it in creating something new. Piet Houtenbas did not intend to make a political statement about the use of grenades; he simply saw the grenade as a beautiful object in its own right, without any reinterpreted meaning.

Grenade Oil Lamps
Designed by Piet Houtenbos. These refillable oil lamps are reclaimed, surplus grenades gilded to add a new esthetic. The grenades transform a somber symbol of war into an illuminating table decoration.

Deer Head
Designed by Augustin Scott
de Martinville for Vlaemsch().
A light and abstract version
of the trophy head, in triplicate.
The moose, deer, and roe deer
heads are all flatpacked for
ease of shipping, and are
easy to assemble.

Bedside Gun lamp
Designed by Philippe Starck.
As part of a series of gun
lamps by Flos, the bedside
piece takes shape in the form
of a replica Beretta hand-held
pistol with an 18-carot gold-
plated finish. The shade has
a silkscreen print internal with
contrasting black diffuser.

A new meaning: design as a statement 59

Above: Wood-becomes-paper (papercup series) Designed by Jeremiah Tesolin. The image on each cup illustrates a different part of the process of making paper. Transparency in production and sourcing methods help informed decision-making on the part of the end user. Photo by Olli Karttunen.

Right: Wood-becomes-paper (recycling box series) Designed by Jeremiah Tesolin. Effective and vivid graphics give the impression of a heavy wooden box and play with the idea of a wastebasket: the woodprint relates to paper and recycling. Photo by Chris Bolton.

Anatomy

The visual identity of a product is comprised of several facets which, together, give it a tangible form. These facets are engineered, to a greater or lesser degree of visibility, through the process of design. This process varies from designer to designer. There are two main areas of consideration—tangible and intangible qualities—in terms of the customer's requirements and the designer's interpretation of these requirements.

The principles of form and function have been well documented. They relate to the tangible elements of a product, with selection of materials and construction methods, ergonomics, handling qualities, and durability based on the desired properties of the product.

The process of designing a product involves more than simply the transfer of information from one medium to another, namely designer to client. It is important to address various ways to facilitate a flow of information from client to end user, including all the departments of the supply chain. In the realms of creativity, do we need frameworks, methods, tools, and information flowing in all directions? During the Memphis period (1981–1988), we saw structures and shapes that blurred the boundaries between art and product. The principle behind this movement was to break all the rules.

Designers need to understand all facets of production in the process of achieving a desired effect in order to generate an exchange of economies. Economies usually mean financial rewards, but sometimes this is not the sole goal of the product's life.

Within product design there are several issues to consider in regard to what creates a desirable product. A designer needs to consider the origination of the initial idea. What is the goal for the project? What is the customer's strategic objective?

There are no rules, set methodologies, or structures in the process of design. The examples of products in this book reveal many means of idea generation and methods by which designers and companies have realized their final product. Design should be open to different ways of working.

Designing for change is the main emphasis. No matter how a designer approaches their work, there is a need for a certain degree of understanding of how other specialists work; how they use tools, frameworks, and methodologies to achieve their own goals within the process of product development. The designer needs the ability to evaluate situations and problems and to critically appraise them in order to reveal a solution. As Jasper Morrison said, the formal appearance of an object could be the visual consequence of an idea, a process, a material, a function, or a feeling. It can also be borrowed or stolen.

Chairwood
Designed by Gudrun Lilja
Gunnlaugsdottir. Another in
Gunnlaugsdottir's series of
"camouflage" chairs.
(See also page 41.)

Strategic objective

Design as a core part of business strategy has never before been the subject of such focus from traditional business theories and methods for best practice. Governments actively seek to promote design within business, particularly in the export of products throughout an international marketplace. As natural resources become a less viable long-term generator of revenue, the world economies look to different ways of replacing and building wealth. The race is on.

The idea of design as an influential and integrated part of business practice has long been awaited, yet only adopted and practiced by entrepreneurial companies. The business of design requires dynamic and flexible ways of working, which are often flattened by company structures.

Information and communication developments fuel the ability to work effectively between specialist areas within supply. New advantages from unusual partnerships or bonds reveal less obvious working methods to develop unique products for the end user. Teamworking; managing change; empowerment; designing the unexpected; brainstorming; customer-centric thinking; activity-based costing; rapid prototyping; long-term views; ecological planning; recycle and life cycle—what does this all mean to designers?

An understanding of, and effective communication with others can uncover informed benefits that could be applied to different stages in the design process.

To work with different producers and manufacturers, a designer needs to be able to see their positioning within several contexts that relate the company to the market and vice versa. Positioning and core company competencies need to be identified by the client (a consultancy, producer, or manufacturer) and conveyed to the designer.

A designer often works within a design strategy for diversification. Diversification may result from in-depth competitor research and positioning strategies that identify market or product extension possibilities. Product extensions might be directly related to existing products, or to unrelated and completely new products within a company portfolio. Redesign or new design?

Adopt new ways of thinking and integrate multiple disciplines through a holistic and empowering design process. Design is not the only part of the process of creating a product; it needs to be seen holistically, as a facilitator for improvement throughout supply.

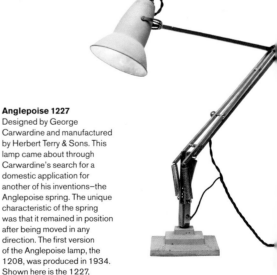

Anglepoise 1227
Designed by George Carwardine and manufactured by Herbert Terry & Sons. This lamp came about through Carwardine's search for a domestic application for another of his inventions—the Anglepoise spring. The unique characteristic of the spring was that it remained in position after being moved in any direction. The first version of the Anglepoise lamp, the 1208, was produced in 1934. Shown here is the 1227, produced from 1938–1969.

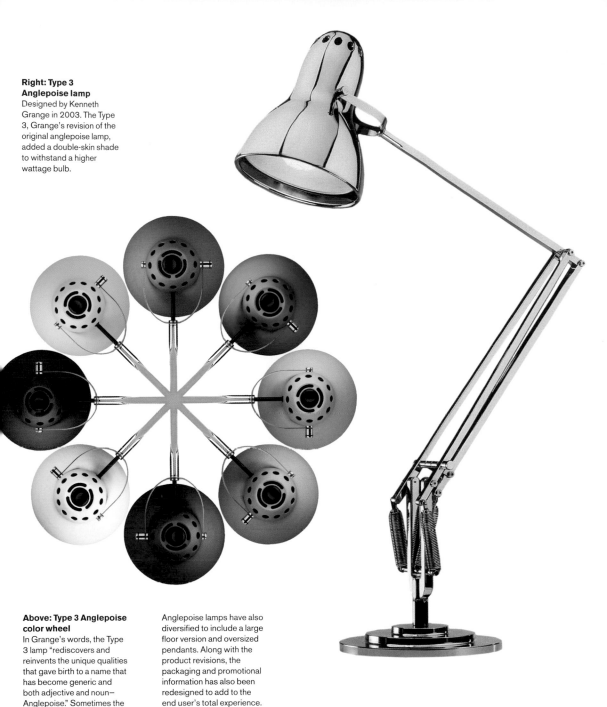

Right: Type 3 Anglepoise lamp
Designed by Kenneth Grange in 2003. The Type 3, Grange's revision of the original anglepoise lamp, added a double-skin shade to withstand a higher wattage bulb.

Above: Type 3 Anglepoise color wheel
In Grange's words, the Type 3 lamp "rediscovers and reinvents the unique qualities that gave birth to a name that has become generic and both adjective and noun—Anglepoise." Sometimes the name of a product can become synonymous with its function. Now available in many colors, designs for

Anglepoise lamps have also diversified to include a large floor version and oversized pendants. Along with the product revisions, the packaging and promotional information has also been redesigned to add to the end user's total experience.

Life cycles

Why do some products last within our societies for so long while others are truly ephemeral? There are two areas to consider: a design idea, and a tangible, individual product life. Design idea relates to those pieces that reach an iconic status and can therefore withstand the test of time, but might also be redesigned slightly so as to maintain contemporary credibility.

Occasionally a product never changes its physical form and is replicated over and over. Examples of products that we think have stood this test of time can be found on the timeline chart.

A product with an ephemeral nature will have a short life cycle due to factors such as fashion and technological obsolescence. Waste is a key, product-specific concern. There are two main types of waste: physical and perceptual. Perceptual waste is caused by cultural forces that determine value and desire; physical waste relates to tangible material waste—direct or indirect residues from a particular process and/or use.

Life cycles for the objects with which we choose to adorn and embellish our lives depend on many factors. If a product has a short life span, how can it be made so as to create less waste? Can we even cultivate this principle of rethinking waste and design a new methodology?

Life-cycle studies have been a part of business methods and strategies for a long time. The traditional life-cycle model relates sales to introduction, growth, maturity, and decline stages; maybe we could add a "new" section to this model to allow for waste as raw materials within the decline stage. First we need to identify why a product becomes out-of-date, then we need to readdress these life span issues. If a product is short-term, redesign and reconstitution should be engineered into its early stages of development. There is nothing wrong with these more ephemeral products, but we do need to think about how they are made and how they can be reconstituted after their initial life cycle draws to a close. Life cycles and their stages need to be broken down and understood by their link to desirability and obsolescence, and from the early stages of design development, we need to rethink life-cycle models and engineer these new meanings into reality and not mere models.

Left: Tall planting pot
Designed by Johannes
Norlander. Made from cast-
molded polyurethane, which
protects plants from both
heat and cold, and which
can be folded down to
form an umbrella stand.
Polyurethane also allows the
planting pot to be recycled.
Photo Mathias Nero.

Right: Magaseat
Designed by Jeremiah Tesolin.
Magazines are folded and
combined to create this stool.
Do all things need to be
altered in drastic ways for
reuse? Here the magazines
are not far from their original
life, and yet offer support
through their hidden strength
and beauty.

Brief

A brief identifies predetermined guidelines and sets the agenda for the design process of a project within the context of production, marketing, and accounting. The designer is often involved in pinpointing certain criteria in the initial stages of drawing up the outline of the brief. Often, the best method of producing a well-defined brief involves several specialized, core departments and personnel working in a small group that can determine feasibility guides for a certain design strategy, even before the writing of the brief.

The objective of the brief is both to inspire the designer and to assist in the realization of the final product. A brief needs to be detailed when producing high-tech products, yet in other projects more freedom of expression is required, so a brief's main incentive is to inspire fresh creativity.

A brief should be a tool that assists the designer or designers with the identification of certain prioritized esthetics pertinent to the project as a whole. Parameters—from cost to constraints of a brand or practical considerations—need to be identified so that a realizable end product that meets predetermined criteria can be created. Where there are restrictions on design choices, for example, due to material tolerances, technical guides are crucial to the designer. Areas usually considered within the frame-work of a brief are key words or a client/brand story that give a linguistic visual for the overall esthetic and emotional context of the project.

Marketing material is also sometimes used to help explain the physical and cultural priorities of the target market. Consideration must often be given to anthropometrics, product logistics, or the ease of transport in relation to value. Accounting information may be included to assist in understanding the final price for a product in comparison to its competitors, or to identify areas of maximum and minimum spend within the production process.

In general, a design brief is a vehicle of communication between the client, which could be a design company, producer, manufacturer, or end user, and the designer. The final brief document can vary enormously; from a simple, descriptive guide to a highly detailed, even scientific, descriptor of calculated standards.

Not all products are created to a brief. Some designers refuse to work to a brief while others enjoy the tensions of creativity within a restrictive brief. More established designers tend not to work to a set brief, but to create their own parameters for each unique project.

Solar Lampion

Designed by Damian O'Sullivan. O'Sullivan set himself a brief to design a light utilizing solar technology in which the solar cells would become an integral part of the design. The resulting light is formed from injection-molded crowns with six inclined solar cells; each of the crowns is displaced by 30°, which results in an organic, illuminated structure. The light is finished with a handle so that it can be moved and positioned easily. Photo by Frans Feijen.

Origination: idea generation and design development

So what is your inspiration? What influences your work? A designer is regularly asked these questions. This area, as part of the overall design development stage, is the main field of interest for providing an insight into the designer and their processes, particularly when a product moves into the promotion and distribution areas of supply. Magazines feature interviews that ultimately attempt to uncover something unique that offers stand-out from other commentaries on design. There is a desire to quantify the imagination of the designer, to understand and see a product from their perspective.

Marcel Wanders refers to "love" as the main influence, both internally and externally, for his work.

Shin Azumi finds relevance in human observation and links present and future. "Daily life. I observe the details of everyday life of people. I think the future is not somewhere in the air, levitating, but on the horizon of the ground where we are standing."

Martí Guixé identifies this need for everyday understanding, but also describes design development with a new term he has coined—Mandorla. "I don't believe in inspiration. I think to create and design a product is about acquiring, developing, and processing information. In my way of working, my material is the information." The main influence on Guixé's Mandorla formula was the ability of a product to communicate, to provide a mystical interface between itself and the user that goes beyond the tangible facets which make up the product's characteristics. The idea behind the Mandorla principle was to identify a formula that could be applied to the design of any product type.

It becomes apparent that there is, to a lesser or greater degree, a need for an instinctive method of layering and uncovering ideas within the early stages of the design process. These ideas evolve from careful research and an understanding of real observations, but also seem to embrace enlightenment where the idea is translated through a deep and personal context.

Idea and concept origination is project specific and follows no set methodology, although there is a need for deep, provocative thought followed by idea reductionism.

Marcel Wanders further elucidates his process. "First, I start thinking, then I create a huge mess. I see tons of little juggling balls, I start throwing them up and see whether the balls together make an interesting solution. If not, I take out the balls and put inside the game new balls only. When the game is perfect, I will be totally happy. Only then do I stop thinking."

Identifying abstract parameters assists in the creation of a unique product, yet the development of ideas and concepts needs also to be grounded in the marketplace.

Sources for reference are extensive within the globally connected, informed world. Resources for primary or secondary research need to be determined in the context of each project, but a good grounding in research to find a unique, yet real perspective is an advantage. Reference material is easily available in many forms of electronic media and printed matter, yet the best resource of all is to be able to improve and create proximity to end users in order to clarify real needs and priorities. Material libraries, trend Web sites, think tanks, Web logs, etc, all add to the information stew. It is therefore vital to analyze reference material constructively.

Maybe we are trying to uncover the anatomy of a designer. Gut instinct does play a role in the generation of ideas, but there are some general principles that can be used to assist in the design process. A designer needs to be able to access, fully understand, and interpret the identified needs of the client, competitive marketplace, and end user in order to engineer the findings into improved product value, while simultaneously finding their own abstract interpretation of this dynamic learning process.

Flowmaker™
Designed and produced by WEmake. This inspirational design tool takes the form of a pack of cards. Photo by WEmake.

Flowmaker

Flowmaker cards provide designers with a tool for inspiration. There are many direct uses for the cards within the design process, yet the main observational benefit is in the use of the cards as a tool to help evaluate and communicate core design imperatives between the members within design teams and the client. Flowmaker, a tool created by the WEmake design studio, offers a method of design discourse that facilitates this vital communication between parties for a given project. It is not a prescriptive methodology, but rather an open-ended, multipurpose, adaptable aid intended to support and extend design processes.

Flowmaker is designed as a pack of cards with only a few pointers for use. It is an invaluable tool at all stages of design—to stimulate, inform, remind, reinforce, nudge, jog, probe, challenge, and inspire. As can be seen from the diagram, there are many areas within the design process in which the cards can be applied in order to clarify or uncover expected or unexpected facets within design. The 58 Flowmaker cards explore instinct, personality, ageing, personality, and potential, represented in five suites.

WEmake are forging the way for design theorists by devising dynamic tools that can be applied to many design-oriented situations. These not only assist designers, but also offer more traditional, static business structures the freedom to explore and reveal innovative design strategies.

Other contemporary design tools also exist, and these should be researched further to provide a balanced and informed knowledge of all the processes on offer. Ideo produced their Method cards; Designersblock produced the Riskitbiscuit model; and the first cards to be applied to design were Brian Eno's Oblique Strategies cards, which were published in the 1970s.

Flowmaker suites
Designed by WEmake.
The Flowmaker cards
provide inspiration and
information for designers.
The 58 cards are allocated
to five different suites, based
on particular areas for
consideration in design.

Instinct	Design to fulfill needs	Feed Fight Flight Nest Philosophize Sex Social – Family Social – Friends Social – Partner	Understanding what motivates and putting the users' needs at the center of a design process helps make a design relevant and easy to integrate into a user's life.
Personality	Design for others	Active – Receptive Daring – Cautious Language – Numbers Solo Player – Team player Specialist – Generalist Traditionalist – Futurist	Looking at personality type polarities helps us to put our own inclinations to one side and acknowledge different mental attitudes and behaviors.
Ageing	Design for our future selves	Eyesight Fine Motor Flexibility Gross Motor Hearing Learning Memory Smell / Taste Speed Strength	We can make mass-manufactured products and public projects easier to use by designing to include those who are less able, but we can also look at making things harder to use, offering up challenges that promote exercise to maintain healthy function into later life.
Play	Design for joy and interaction	Chance Copying Discovery Head Spin Head-to-Head Self-expression	Designing interactions that absorb us and elevate our emotions can lead to more engaging, enjoyable user experiences and help promote mental, social, and physical fitness.
Potential	Design for sustainability	Adapt Customize Cycle Dematerialize Empower Locate Maximize Meaning Mend Self-make Upgrade	Making the most of the elements used, the people involved, and the context in which the project sits in order to extend the material and experiential possibilities in this and future lives of the design.

Exercise can help maintain
the muscles needed for
balance and co-ordination
and reduce the risk of falls.

BALANCE

Choosing to surrender to
uncontrolable events.

CHANCE

Is an object necessary?
Can you use something
that already exists?
Is your design a service?

DEMATERIALIZE

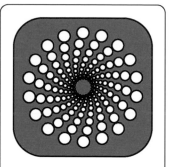

Dizziness,
light-headedness,
rush, vertigo.

HEAD SPIN

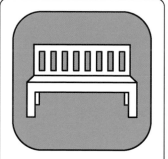

Listener, follower,
spectator,
open, sensitive.

RECEPTIVE

Alliances, empathy,
peers, contemporaries,
commonality, communication,
mutual support.

SOCIAL - FRIENDS

Responding to danger
or discomfort by
running away, escaping
[physically or mentally].

FLIGHT

Jill / Jack of all trades, pursues
a broad range of activities that
have a low entry threshold.

GENERALIST

Aerobic, muscular and skeletal
strength decrease with age.
But strength can be increased
at any age through exercise.

STRENGTH

Can the design be upgraded
further down the line? Do
different parts of the design
have different lifecycles?

UPGRADE

Flowmaker cards
Designed and produced
by WEmake. There are no
rules for using the cards;
either select from particular
suits or randomly from the
pack. Use them as design
tools to stimulate inspiration
or to provide a framework for
brainstorming and planning.
Shown here are 10 of the 58
individual cards.

Designing innovation

In a globalized era, where design seems to be evermore homogenous and high-tech industries share research and development to reduce costs, designers need to work harder to create more distinctive products. This stand-out can be achieved by reworking a product, giving new value to a latent life, adding functionality, or creating something radically new and genuinely innovative.

To innovate and create change can be difficult, as with age and experience also come restrictive methods or ways of working. To be innovative requires a dynamic outlook that keeps morphing and embracing new movements within technological and sociopolitical environments. Juxtapositions reveal tensions: commercial pressures versus personal creativity; free growth versus cultivation; convergence versus divergence; collaboration versus a solitary approach.

Unforeseen and even strange observations may become fruitful design leads upon research; designers can find counterintuitive solutions through product development. To design innovatively is not to expect a "eureka!" moment, but to uncover a novel design idea from a holistic and inclusive design process. This requires dedication and hard work. The designer must forge through the preconceptions of what design is or should be and create a personal perspective; a unique personality and style.

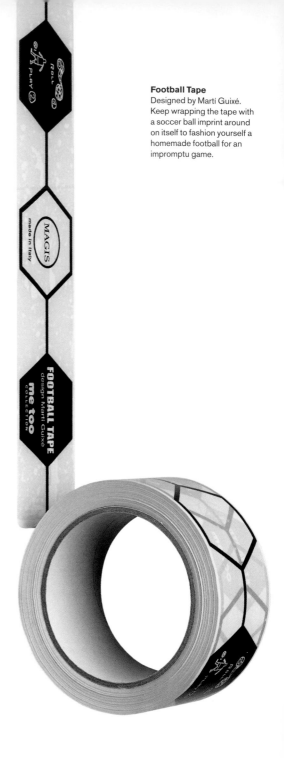

Football Tape
Designed by Martí Guixé.
Keep wrapping the tape with a soccer ball imprint around on itself to fashion yourself a homemade football for an impromptu game.

Toaster Fun
Designed by Jason Miller.
Experimentation can provide
interesting results. Jason
Miller uses what appears to
be a traditional toaster
to embellish toast with
decorative symbols. Visually,
it is not the toaster that is
redesigned, but the toast.

Pixel Tape and Instant Labeling Tape
Designed by rAndom International. Instant Labeling Tape is made up of a sequence of 14 segment displays that can be blacked out with a marker pen to create labels, installations, or signs. Customization is designed into both the Instant Labeling Tape and Pixel Tape through the interactivity of the end user and tape, which hands over control of the final graphic image to the user.

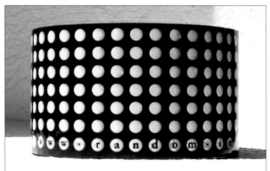

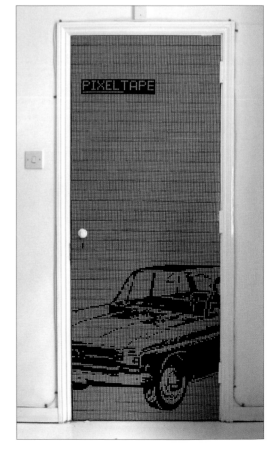

Designed by Gudrun Lilja Gunnlaugsdottir for StudioBility. StudioBility's Visual Inner Structure series gives old chairs a second life. Gunnlaugsdottir strips back the old chairs then covers their frames, including the springs, with felted wool.

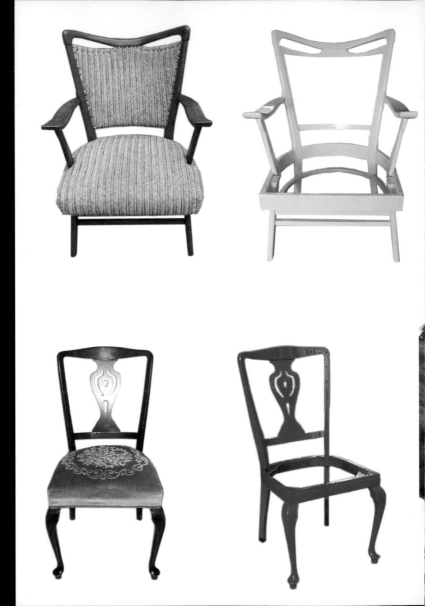

Sticks

Designed by Hsu-Li Teo and
Stefan Kaiser. An innovative
screen design with glass fiber
rods that slot into a recycled
rubber, cork, or timber base.
The silhouette effect is as
inspiring as the product itself.

Product development process

The key areas for the design and development of products are: research, planning, innovation and idea generation (all discussed in previous chapters), communication and concept development, production development, and supply.

There is an ever-increasing focus on customer-centric design and, through research into the social and economic position of end users, a product can find greater market relevance. The idea of market relevance is to identify and determine perceptions of value and how best to prioritize these observations within the context of the product/end user/marketplace. There is a need to find a balance between the findings and the company's core competencies and to clarify their positioning in a competitive product world. The designer must access this knowledge and become an integral part of the company, find a symbiosis between departments, and help uncover potential unique product positioning.

Once product placement is identified and a concept has emerged, the product needs to be defined; team members must work together to clarify and pare down the product characteristics. Every aspect must be considered and approved by the designer, key specialists, and the client, with 2-D and 3-D modeling. This part of the development process examines the design in terms of components relating to the overall look and function of the product, and results in a final model providing a defined insight into the end result.

Sittingobjects: Wound Up, Wound Middle, and Wound Down
Designed by Yara den Hertog. Hertog wanted to create a "volume" to sit on. Experiments with volumizing materials and methods resulted in these visually intriguing stools fashioned from steel ribbon. Working closely with the steel ribbon manufacturer enabled Hertog to develop the Sittingobjects series—a good relationship between production and design is essential in research and development.

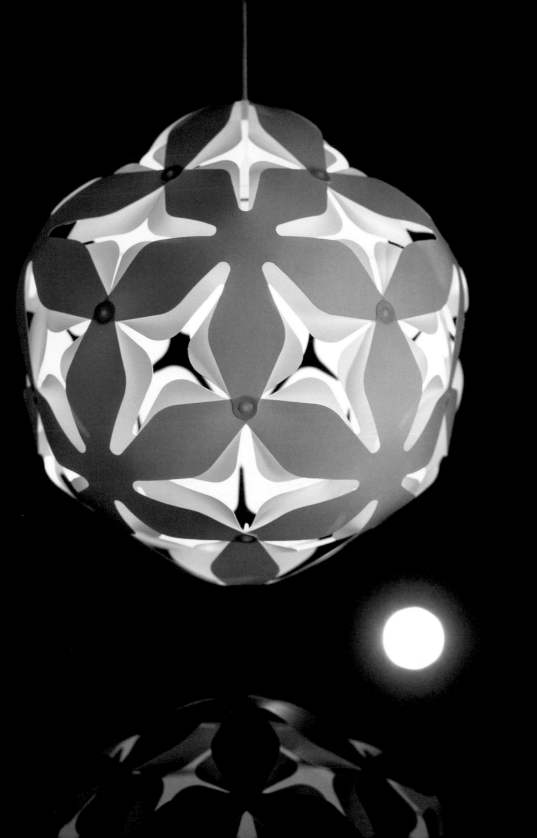

The designer must then test and translate the models and drawings from concept and design development into a tangible product. Production specifications provide the information for the realization of the original idea. This is shown through detailed specifications and the tooling necessary for running the production methods. At this stage, designs can still be revised and withdrawn to be reworked. As this tends to involve higher costs, it is more resource-efficient to revise and clarify potential problems within the earlier stages of design and development. It is the ability to work as a collective or team and to experience an ease of information and knowledge flow between involved parties that prevents expensive redesigns within the stages of real production.

So when does a designer hand over a project? Some may think a designer's job ends here. But the process of product development extends from the area of origination, planning, design, and production development through to influencing areas of distribution, logistics, and promotion. No part of the process team works in isolation, and the best product solution is only achievable by employing a more holistic approach.

A designer is a member of a team and can help guide that team through design and product development. Within this team there should be a good cross section of people who have experience and knowledge of their specialist areas, and dialog between all team members is vital at all stages. The degree of knowledge each can impart at different stages of the process will vary. Some will have more or less influence at these different stages, yet all need to be part of the production of the definitions, especially during research and planning, in order to reduce the potential for product failings.

Pollen Light
Designed by two create for 2pm. Revision and development of the shape through experimentation with material and structural form resulted in this sculptural pendant light. Paper models created in the two create design studio(right). Photo by Jake Curtis (left).

Tools for the designer

The tools of a designer can take many forms, both physical and perceptual. Tools for the development of design rather than the production of the end item vary from the paintbrush to the computer, yet they all share a common factor—they are the means by which a designer defines an idea internally and communicates these findings to others. Tools include diverse collections of objects and equipment that will hold a greater or lesser value to each individual within the context of a particular project. Tools can be objects that fuel the generation and definition of the original idea, through to the hardware that maps its shape and details its production: collections of materials or finishes, objects lost or found, libraries of magazines and books, scrapbooks and storyboards, cameras, computers, software programs, printers, a ruler, a scalpel, a notebook, a pencil …

In contemporary design, giving visual form to an object will more often than not involve the computer and various design packages, such as 3-D Studio Max or Photoshop. 3-D Studio is just one of the modeling programs that has speeded up the entire design process. It allows designers to quickly render a 3-D image of an idea, which enables them to tell immediately if an idea can work visually without going through the lengthy process of modeling by hand. Applications such as these also open up the world of design to those with ideas who may not have the creative methods of expression.

There are advantages to the increased pace of creativity heightened by the use of CAD systems. A designer can produce a greater number of ideas within shorter timescales. As improvements continue, will the client expect so much more from designers? Will design actually integrate further and become manufacture. This is already happening with the blurring of design and manufacture through intelligent production methods such as rapid prototyping. Throughout all stages of the design and development process there are tools that benefit the production of quality products; tools that are instrumental within each stage may become obsolete as the next new product progresses through to supply.

The switch from traditional craft to a new, mechanized design esthetic has allowed for an interesting cross-pollination of craft-type detailing that can now be applied to objects in minutes instead of hours. This switch to machines as designers' tools speeds up the production process and removes the need for a large workforce of craftsmen. But, with this in mind, have some of the objects around us lost a certain human, hand-crafted quality?

Double Pen
Designed by Sam Hecht. Designed as part of a series of six personal items for Lexon, this pen offers the user the choice of rollerball or a gel pen.

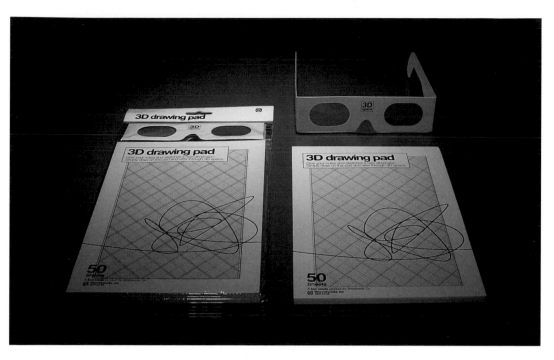

3D Drawing Pad
Designed by two create.
A sketch is often the first
visual form a design idea will
take. Ideas often occur in
unlikely places, but with the
3D Drawing Pad you can be
prepared at all times, and
visualize your ideas in 3-D.

Sketching and drafting

Most designers begin the design process with a pen or pencil and paper. There is a beauty and a freedom to sketching by hand. There are varying degrees of ability and drawing styles that make this process very personal and distinct to each designer.

The application of sketching has a temporary nature, where marks can be revised and redrawn until the preferred effect becomes defined. Tools of a much more ephemeral nature are often the first visual representation of a thought. The pencil was invented in 1560 and has changed little since then, except for the replacement of lead with graphite. A pencil assists thinking and creativity as it allows us to erase, redraw, and develop thoughts into drafts and plans.

For designers BarberOsgerby, pencil and paper are the starting point for any concept. "Sketching is a vital tool for any designer as it allows you to develop your ideas quickly. This is always the process in which ideas are conceived." Sketching and drafting fulfils many objectives: informative, esthetic, artistic expressions; shape-forming, and technical detailing …

Drafting is a specialist skill that, traditionally, was executed on paper, but is now most often performed via CAD and digital drawings. Information relating to material tolerances and shape specifications for the purpose of product or component production can be developed quickly, accurately, and collaboratively on screen.

Drafting, whether on paper or screen, reveals the form of the object, the basis for the transformation of conceptual brainchild into real product.

PH 3/3 Lamp

Designed by Poul
Henningsen, drawing by
Ib Anderson, 1927.
Henningsen originally studied
architecture, and designed
his first chandelier as part of
an interior design exercise.
Designing lamps became his
passion and his work revealed
something of his character
with its interplay between
reason and feeling. The
drawings provide a link to
the designer and reveal the
unique style that underpins
all his work, which juxtaposes
art and technique, images
and thought, light and material.
This drawing is an effective
and detailed representation
of a design that has become
a trademark for Denmark.

Modeling: rapid prototyping

Within the process of design there are stages that translate ideas into varying degrees of reality. Designers each have their own unique approach to this area of concept visualization; model-making is a tool within the process that offers either a hands-on or a computerized methodology. Creating a handmade model involves a skilled process using cardboard, foam, or whatever works for testing and analyzing the design. Modeling performs a dual function: to present ideas to customers and, as Sam Hecht of Industrial Facility notes, to achieve a clear and often unexpected result. Modeling is an important part of design as a process for reducing, evolving, and perfecting the technical and esthetic quality of an idea. Sometimes a few models are created while other times many iterations are required to achieve the final form. Many designers use this hands-on approach, while others are moving toward more computer-generated options.

Technological developments in both materials and processes have a profound effect on designers and the products they create. In many cases, such developments are borne of the race for new research and production methods in the space, car, and medical industries.

Rapid prototyping is used for the modeling of components that require low tolerances in fabrication, that is, that require absolute precision engineering. The time it takes to create prototypes that might then need quick

Snotty Vases
Designed by Marcel Wanders. Every sneeze is different, as illustrated by the complex detailing of these vases, enabled by rapid prototyping. These images illustrate the process of rapid prototyping.

THE MODIFICATION OF THE DIGITAL SNOTTY

OZRENA

ADVANCED TAN SOFTWARE PROGRAM WHICH TRANSLATES CAPTURED ORGANIC SUBSTANCES INTO COMPUTERMODELS

MODEL-EDITING AND CREATING SPECIAL DESIGN SOLUTIONS

THE GENERATED COMPUTERMESH WILL BE SENT TO THE SLS-MACHINE TO PRODUCE A UNIQUE MODEL OF THE CAPTURED PARTICLE

OZRENA

OZRENA GENERATED

THE PRODUCTION OF THE FINAL 3D OBJECT

THIS MACHINE PRODUCES THE FINAL OBJECT USING THE 3D DRAWING. THE MODEL IS BUILD UP BY THE COMPUTER GUIDED LASER.

A DIGITAL 3D NANO-SCAN OF A SNOTTY

PROTOTYPE OF A NEW MICRO-TECH ADVANCED SCANNING-DEVICE. ESPECIALLY ENGINEERED TO SCAN ON MICRO-TECH BASED LEVEL

EQUIPPED WITH OPTICAL NANOLENSES

POWER SWITCH

OZRENA PATIENT PREPARING TO SNEEZE INTO THE SCANNING-DEVICE

OUT OF THE WHOLE 'RAIN' OF PARTICLES ONE PARTICLE IS SELECTED AND TRANSFERRED BY A 9 BIT/RATE PROCESSOR TO A COMPUTER DEVICE.

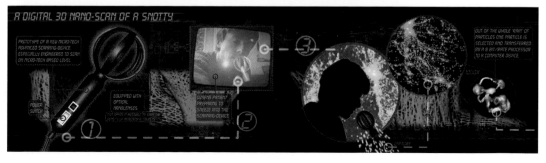

modifications is greatly reduced. This merging of computer-aided design and manufacturing processes results in a greater efficiency for product design.

Rapid prototyping allows for finer and more detailed intricacies of design, as the process layers material into a 3-D version of the original idea. It is this fine layering that allows for complexity, and each layer is so thin that curvature and fluidity within the realm of sculpture translates into a more complex product. This process was invented in the mid-1980s and since then has become an ever-developing application for designers in both model and batch-production of products. The benefits of rapid prototyping are seen in the freedom to create new shapes, the reduction in time and expense through increased efficiencies of model-making within the design process, and a reduction of production costs. Also, with rapid manufacture, we have the opportunity to steer away from mass-production due to the ability to offer smaller batch, customized, and one-off production pieces.

Specialist skills—including the ability to use 3-D software and to translate sketches into a precise model—are required for the application of rapid prototyping and manufacturing methods. A knowledge of these applications enables the user to shape exacting mathematical structures. The designer does not necessarily need the skills for this process, as this part of the modeling process can be subcontracted out to pattern and model-making specialists. A designer can provide specialists with a sketch or drawing on paper which can then be synthesized into a fully formed product.

There is always a definite need for the ability to apply varying degrees of the skilled, hands-on, workshop-style approach to the fine-tuning of ideas and concepts for products. As well-designed, more complex, and even customized products are demanded, designers will develop shortcuts and new ways of working. This new process of production for one-offs or complete customization is on the way to becoming the catalyst for true customer-centricity.

Osorom
Designed by Konstantin Grcic for Moroso. Formed from a new, blended plastic made from fiberglass and resin, this seat plays with perceptions of solidity and lightness.

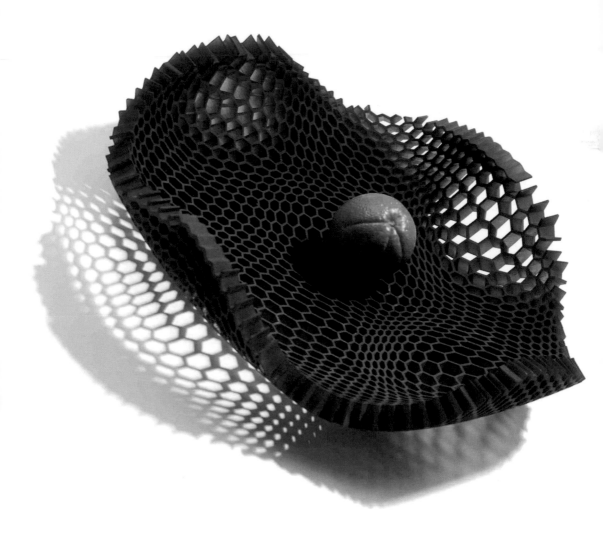

Black Honey.MGX

Designed by Arik Levy. Influenced by the geometric shapes repeating in the structure of honey, this vessel captures a fluid effect with the use of rapid prototyping technologies. The fruit bowl seems to have grown organically into a detailed and irregular, but naturally occurring shape. Such a detailed structure could not be produced with conventional methods of production; the process of rapid prototyping enables the sequential layering of material to build up the predetermined final shape.

Communication and prototyping

Prototypes should be utilized as a means of developing the right concept and facilitating informed decision-making that will ultimately avoid lengthy and costly development and production timescales.

Presently, much time is often spent in the area of creating visual models with CAD systems before prototyping; yet there is a need to view prototypes in great detail as a means of getting the final product right. The number of prototypes created varies enormously, depending on the scale and budget of a project—from none to thousands! The level of intricacies in shape, composition, materials, and production methods, from conceptualization through to the final model, affects the number of prototypes produced. Prototypes are vehicles of visualization that prompt dialog and facilitate fine-tuning in order to harmonize the product characteristics and uncover any potential problems within the production process.

Prototypes, as vehicles for communication, provide all team members with a tangible means with which to validate the product before it goes into production. Prototypes assist in the resolution of a product's potential to make it through production, marketing, and distribution, and to create revenue. A visual identity is engineered and the validity of certain characteristics can be discussed.

As many designers are freelance and contractual, and are often not an integral part of a company, trust needs to be gained and knowledge-flow facilitated as much as possible to ensure the best product solution. Often designers work for a client within their own collectives and teams, but the need for models and prototypes is still a vital tool for dialog. Once a product has been proved to be successful through the testing of the prototype, then it is likely that the designer will be employed again and again by the same company, as teams are built on trust and understanding when it comes to developing "new" products.

Surface Chair
Designed by Tobie Snowdowne at two create for MDF Italia. Made from laser-cut sheet aluminum finished with carpet. Good communication between designer and producer allow the finer details such as finishes and color choices to be fully evaluated before selection. Prototypes of the Surface Chair are shown here. Photo by Jake Curtis (right).

Materials

The selection of materials for a given project is influenced by physical properties relating to durability, sustainability, and shape-forming—the study of which is based in specialist areas of ergonomics—and is important to the esthetic characteristics prescribed by the designer and end user.

Historical developments in material technology have forged the way for product design. The first steel tube was bent into a chair in the early 1990s. Industrial methods of production and computerization had a huge impact on the man-made world, and materials and methods of production are being continually redefined. Synthetic materials and hi-tech production methods coexist with natural materials forged by ancient craft methods—but with obvious time and cost implications.

Material developments usually evolve from the scientific creation of something unique, or from reapplying or reusing past materials that might have been forgotten or have gone out of fashion. Material development provides opportunities for the designer to create the ultimate—to offer something that has never been seen before.

The Miura Stacking Stool, made from reinforced, injection-molded polypropylene compound, is robust yet also offers a dynamic and colorful form that shows off the marriage between construction and ergonomics. This element of surprise, or new application of old materials, highlights a smart approach to design.

Materials not only affect the design process; they can also engender different emotional responses. Interesting material applications offer great shapes with tactile references. The Surface Chair, upholstered in carpet, exemplifies this approach; it resulted from the study of sitting positions on the floor. In it, the carpet is extended from the floor covering onto a chair, thus offering a more comfortable and easily maintained product. Material selection is key to creating mystery and ambiguity.

The friendly shape of the resin-reinforced chalk speaker on page 104 gives a tactile smoothness that demands to be squeezed. Material selection matched with wireless functionality shows off the concept of conspicuous minimalism.

It is not only new materials that can offer a route to design innovation: traditional processes, crafted techniques, or old (even recycled) materials can be imbued with new life through previously unrelated applications.

Miura stool
Designed by Konstantin Grcic. This injection-molded polypropylene stool has a stacking function that saves space and looks good. The materials and production process selected allow an extensive color choice as well as functional stability and pleasing esthetics.

Left: Super Elastica
Designed by Marco Zanuso
Jr. and Giuseppe Roboni
for Vittorio Bonacina.
Established in 1889, Vittorio
Bonacina remains to this
day a specialist, family-run
manufacturer of rattan
furniture—proof that with the
application of craftsmanship,
traditional methods and
materials really can stand
the test of time.

Above: Fresh Fat series
Designed by Tom Dixon. The
Fresh Fat series is a key part
of the Tom Dixon range. He
extrudes lengths of warm
Provista, an amorphous
copolyester resin produced by
Eastman Chemical Company,
which he then weaves, twists,
and molds into extraordinary
bowls and furniture pieces.
The final forms are fixed in air
to form solid structures.

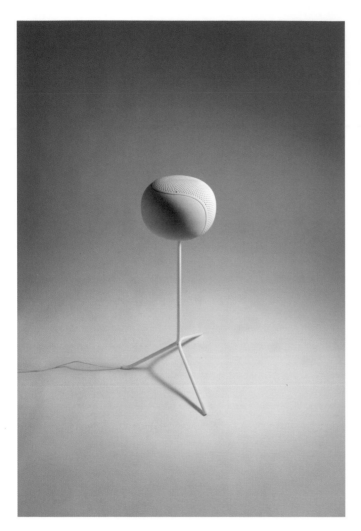

Wireless Speaker
Designed by Kyoko Inoda +
Nils Sveje Architecture.
Resin-reinforced chalk with
perforated detailing and a
monochrome color scheme
give a sleek speaker design.
Music is transmitted to the
speakers via wireless
broadcasting, allowing for
ideal speaker placement.

Above: Paraffin table
Designed by Timo Breumelhof.
Made entirely from paraffin,
this table can also be used as
a candle, with the inevitable
end result being the
destruction of the table.

Right: Inhale vase
Designed by Yara den Hertog.
As the name suggests, this
ceramic and stoneware vase
expresses the process of
inhaling: hollows in one side
result in bulges on the other.
A platinum outer skin gives
the structure strength.

Color

Color is a very specialist area of study and should be taken seriously within product design. Color has a psychological effect on our moods and behavior; it affects how we respond to our surroundings. Color is not purely a visual manifestation—it affects the mind on a subliminal level.

In the 4th century BC, Aristotle noted that blue and yellow references the moon (darkness) and sun (light). In 1672 Newton wrote, in his controversial paper "Opticks," that light can be separated into colors of the spectrum. The Bauhaus educational establishment of the 1920s taught a new approach to the design process and adopted the study of color and emotion along with color and shape. Systems help identify, define, and maintain similarities and differences within the study of color. The Munsell system, developed in 1898 by Albert Henry Munsell, is probably the most widely used method of color notation. Munsell's system modeled an orb containing the full spectrum of colors, which he then described using numerical values of three aspects: hue, value, and chroma.

The selection of a certain color depends on the relevance of the product in use and in the end users' context. Vivid colors create a strong impact that attract and draw attention, but will not necessarily equate to sales. Color can highlight and reinforce, or blur a detail of a product; it can unify and simplify the general form. Seminal French designers Ronan and Erwan Bouroullec, for example, usually apply a monochrome approach to color within their designs or use the naturally occurring color of a material.

Product design needs to take on a greater understanding of the use of color and both the tangible and intangible effects it has on the end user and their purchasing decisions. Color is usually considered toward the end of the design process and is rarely the starting point. Color selection depends on the end use of a product and its target market.

Strap
Designed by NL Architects for Droog Design. Rubber can be easily colored and has good tensile properties. These colorful bands are wall-mounted to form strong bands for storing or displaying objects. Photo by Eric Calvi.

PANTONE®
13-0859 TPX

Left: Pigeon light
Designed by Ed Carpenter.
Made from colored, vac-
formed Perspex. With Pigeon
available in pink, orange, gray,
and white, and the flex
available in pink or white, the
cable becomes an integral
part of the whole design.

Above: Flight Stool
Designed by BarberOsgerby.
A limited edition of signed and
numbered Flight Stools were
made for a pantone color
installation at the ICFF in New
York. The collection, made in
bentwood, consists of eight
different color sets with six
different gradations in each
set. Inspiration came from
Pantone chips and every stool
has the Pantone reference for
its color printed on one side.

Texture

A product's texture can provide a tactile experience or a visual play on its perceived tactility upon contact. The skin of a product is the human/product interface where, practically, the internals are concealed and protected from everyday use, and, experientially, the user perceives sensory responsiveness. There are many ways to induce an intended response through the selection and application of a multitude of materials, finishes, and processes to create the desired properties of the product "skin."

Generally, material selection will involve the consideration of texture as part of product development, yet the recent move toward visual and tactile textures reveal that this property must be considered independently of materials. Texture can be given to a surface in many ways to add to the product experience: printing, embossing, lacquering, painting, polishing, brushing, layering, adding textile structures, and many more variants.

So how can texture affect the interaction between user and product? In a visual and tactile sense, the play of a flattened or enhanced surface provides more information to the end user and can heighten or dampen the user's response to the object's hidden or accentuated attributes. Texture can highlight overall or specific areas of functionality, shape, and decorative effect.

Consider the Apple iPod's smooth, ghost-like appeal; there is minimal interruption as the hand passes over the surface. This is appealing to the physical sense of touch and therefore has the potential to enhance your perception of product functionality and quality. The application of different textures provides the user with information in order to assess and understand the placement of a product within their environment.

Left: Skin beakers
Designed by Royal Tichelaar
Makkum. The unusual textures
of these beakers have been
created by printing them with
materials such as tape, rubber,
thread, and orange net.

Above: Honda Vase
Designed by Erwan and
Ronan Bouroullec. The deep
aubergine or black matte
backdrop of this fiberglass
vase shows your flowers' true
beauty. The matte internal
finish contrasts in texture and
hue with the shiny painted
outer. Photo by Morgane Le
Gall courtesy of Kreo Gallery.

Left: Wind rug
Designed by Paola Lenti.
Known for his research and
experimentation in yarns
and textiles for upholstery
and flooring, Paola Lenti's
products are unique. The
Wind rug, part of his Aqua
collection, is made from Lenti's
signature Rope material. The
thin, round braid, developed
specifically for this piece, is
hand-tufted at a specific
height and density to
reproduce the softness of
a terry cloth fabric. Photo
by Sergio Chimenti.

Right: Biterknife
Designed by Jeremiah Tesolin.
A butter knife decorated with
bite marks. Textured with the
impressions made by teeth
marks, silver and gold inlays
add the finishing touch.

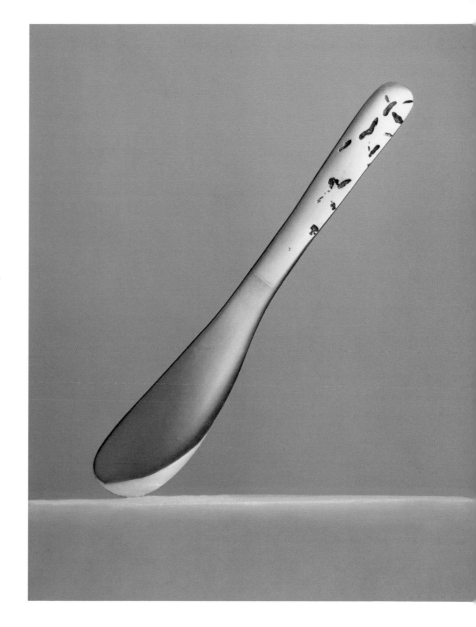

Proportion

The proportion of a product needs to be set in the mind of the observer as a specific object in the context of a particular surrounding or other designs. To question what proportion is and how it affects the design process requires us to look at historical ideas on this subject in relation to geometrical figures and mathematical sequence. Ideas of harmony were applied to the design and construction of the pyramids and the Parthenon, and Leonardo da Vinci applied his principles of "divine proportion" to art. Many have studied the connections between beauty and harmony through mathematics and the principles underlying proportion.

Architecture, natural science, cosmology, sculpture, and drawing have all applied the principle of Phi—the Golden Section. The influence of ideas relating harmony to beauty are dotted throughout history. What are the defining characteristics that add to a harmonious relationship within a design, and how do they relate to each other and the whole? Is this harmony intuitive?

Ideas have been given form through the application of progressive, geometrical sequences. Designs with pleasing relationships between the parts of a whole, and between the whole and its surroundings, require a consideration not only of the actual object, but also of the relationship of the object to its environment. Some designs just seem right, yet when placed within a different context they lose the sense of balance that draws the elements together. In questioning what proportion is, both within a product and between the product and a certain environment, design characteristics become highlighted or subdued in many defining ways. Every individual interprets the placement of products within an environment in their own, personal way; the result can be intrigue or a sense of peace and beauty. A designer questions the recognized "truths" about proportion, and their application, through the design process.

Plant Cup
Designed by Gitta
Gschwendtner. When an
oversized design is placed
within an "unexpected"
setting, careful attention
must be given to their internal
and contextual proportion.

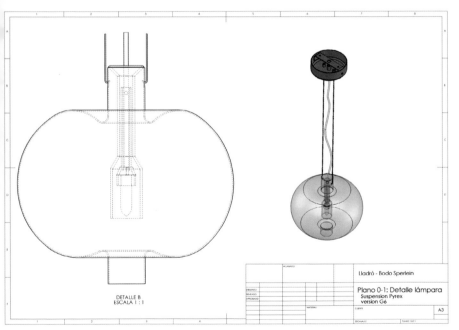

DETALLE B
ESCALA 1 : 1

		RICHMOND		Lladró - Bodo Sperlein
DIBUJADO				Plano 0-1: Detalle lámpara
REVISADO				Suspension Pyrex
APROBADO				version G6
		MATERIAL	CLIENTE	A3
			ESCALA	PLANO 1 DE 1

Above and left: Bird pendant light
Designed by Bodo Sperlein. Handblown glass is decorated with handmade, bone-china birds. With shadows created by the lamp a part of the overall effect, an eye for proportion was vital in the design of this light.

Right: Big Clip
Designed by Arash Kaynama. The paper clip first appeared in 1899. There have been several versions of the paper clip since then, all of which applied a process of bending and shaping a thin rod of metal into a functional shape for the purpose of clipping papers together. With the Big Clip, of course, the function is a little different; it can be used as a magazine or newspaper holder, or a clothes hanger. This version of the clip measures 11 x 3in (28 x 8cm). Photo by Kelly Sant.

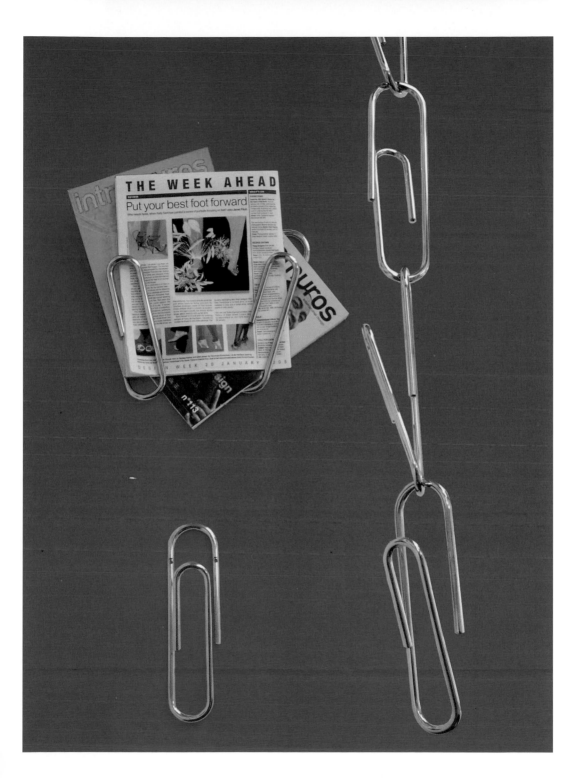

Material process and production

Just as there is a multitude of materials, there are many ways to construct any given product. There is a fundamental richness in the diversity of means of production, allowing the designer a freedom of expression in which work can embrace the ideals of industry, through to those of arts and crafts. Designers learn new skills or methods through working with different companies that specialize in certain production skills or unusual ways of working.

Some companies seem to take a branded approach of artistic expression, producing many types of products in diverse ways. Moooi is creative in an arts-based way, utilizing new materials or processes, or reworking crafts-based, labor-intensive methods of production. Other companies adopt a particular material or process that becomes their industry signature. Magis use methods of injection and rotational molding within many product categories, alongside other process methods in their product portfolio. Inflate design and produce inflatable temporary structures in rip-stop nylon for indoor or outdoor events. Although they specialize in inflatable structures, they also offer dip-molded products within their product portfolio.

Traditional crafts, in which timeless skills are applied within contemporary designs, are found in older, specialist companies like Royal Tichelaar Makkum. Using tried-and-tested methods of production while working closely with designers, Tichelaar find new and refreshing ways to develop their product portfolio.

As designers experience the high-tech principles of production methods used in other industries, their time-compressing effect can be seen in the way products are created; production schedules and time-to-market are reduced. A certain process can quickly become the latest fashion within product design, so designers must beware of the potential for overkill and market saturation of certain processes. A designer builds a knowledge base while experiencing new and traditional skills from interacting with specialists of different processes.

Crochet Lampshade (top) and Knitted Lampshade (bottom)
Designed by Tim Denton and Johanna Van Daalen at electricwig. The lampshades are hand-crocheted or knitted in cotton, and stiffened with a mixture of sugar and water. Photo by Tas Kyprianou.

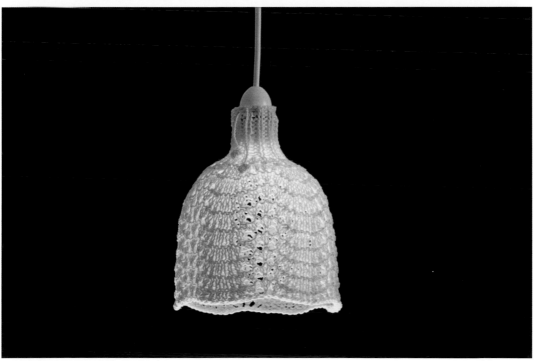

OIAB (Office in a Bucket)
Designed by Nick Crosbie.
This portable, inflatable
meeting space is made from
a lightweight, polyurethane-
coated, ripstop nylon.

Lace Fence: How to Plant a Fence
Designed by Joep Verhoeven at Demakersvan. Why does a fence—a boundary marker—need to be ugly. Lace Fence, made from plastic-coated metal, is both uplifting and space defining. Photo by Raoul Kramer.

Industrialized Wood
Designed by Jeroen
Verhoeven at Demakersvan.
The initial reaction to this
piece is "how did they do it?".
The attention to detail and
craftsmanship of the 17th
and 18th centuries is revived
in this table. Equally exacting
methods of production are
used, but with a twist, as most
of the work is now completed
by machines. Fifty-seven
layers of plywood, cut by a
five-axis CNC mill, are glued
and sanded by hand to create
this astonishing table.

Naming

To name a product is to place it in a new context, to make it familiar, to leave a mark in time, and add another dimension to the way it is viewed. Naming is an important part of the design process and demands consideration in a world in which it is increasingly hard to gain visibility.

Historically, there has been an obvious relationship between names and objects in design; many products have been given animal or place names, for instance: Pelican Chair, Yemen Vase, Papillon Chair, Snoopy Lamp, and Penguin Donkey. Animal references bring a new relevance to the product and its use, through the geographical concept of having the world in your home. A name can convey perceptional qualities: a Diamond Chair must be precious and should be treasured. Often a name relates to the main characteristics of a product's appearance or to a certain function it can offer. In other instances, a name is invented to give the product more futuristic qualities, for example, Nokia and iPod.

The key considerations in naming a product are problems that might arise from direct translation and whether the name is inventive or evokes certain experiences or observations. To give a product a name is a difficult undertaking in a global setting, where linguistic diversity, translation issues, pronunciation, and meaning must all be considered. The name of an object should flow off the tongue and be pleasant to say in as many languages as possible.

Naming can be a powerful means of offering a direct visual image of a product or adding a layer of ambiguity, with an indirect or completely fictional interpretation of the product. A Wednesday Light makes us feel that it is not a Monday or Friday light—it is made at the peak of the week, when the maker is at their best and is not considering the weekend. 1132 Minutes enlightens us with an indirect relevance: it is the time it took someone to physically make the piece.

A good name is a powerful force in branding and can help differentiate or add emotional context to a product to engage your customer. If you can conjure a certain feeling in the user by emphasizing the intangible and mystical qualities that indirectly reveal the product, you have an edge on a less evocatively named competitor. The Essence wine glass relates to the evocative nature of a finely refracted liquid of exquisite quality to refresh every part of your being.

While some may not consider this an important area for the designer, it can directly link the designer, maker, and end user through their subconscious and conscious view of the product.

Left: Essence
Designed by Alfredo Häberli. Häberli wanted to design a glassware collection that had thoughtful balance in both appearance and end use. The result is a collection of wine glasses that capture "the essence between tradition and modernity, between celebration and daily use," and that all have the same balance between stem and base. Photo by Iittala Oy AB.

Above and right: Diamond Chair
Designed by Harry Bertoia. Bertoia created this airlike design in 1952, having experimented with industrial wire rods. The name not only echoes the shape, prompting an immediate visual recognition, it also has the connotations of a highly valued treasure.

Above: Penguin Donkey 2
Designed by Ernest Race in 1963. This was the second design within the Penguin Donkey collection by Isokon. In relation to this product, the name "donkey" suggests the characteristics of functionality and reliability.

Right: Snoopy table lamp
Designed by Achille and Pier Giacomo Castiglioni in 1967. With its solid marble base and enameled reflector, the Snoopy table lamp was named for its beagle-like "nose."

Far right: Pelican Chair
Designed by Finn Juhl. This chair takes on the pelican's form as it stretches its wings before flight. Originally made c. 1940, this design was one of the most shape-challenging pieces of furniture of the period.

Promotion

The items we purchase reveal something about who we are as consumers. The magazines and newspapers we read, our address, our names, and our broader consumer profile can give marketers a means of segmenting us into groups of like-minded individuals. Reversing this observation of marketing methods highlights the importance of the right distribution and promotion for a certain company or product. The magazines, newspapers, shows, galleries, and shops that combine to promote a product will affect the end user's perception of the product, along with the brand and name of the designer.

Products can be unique in that they are one-off or even part of a limited-edition series, or, conversely, mass-produced and available to a specific market. A designer adds layers of perceived value into the product, yet once it leaves the factory doors the product embarks on another part of its life. To give a new product context within our diverse and diffuse distribution channels requires careful planning. The designer needs to be aware of all stages of distribution, as promotion in particular will refer directly back to the conceptual part of the product's life and the designer's image. In the early stages of a designer's career, the amount and quality of press received will help sculpt and forge a channel to a future design career.

In general, there is the saying that any press is good press. While this is partly true, the right kind of press is critical for long-term value. As a designer embarking on the journey of image-making, one should list the magazines, Web sites, Web logs, shops, galleries, and journalists that have an affinity to the product style and type: keep them informed about new product developments, but do not plague them. The methods and style of communication used to keep key people and companies up-to-date is an important area for consideration.

The designer must consider the ultimate goal of a particular product. Is it to get as much press as possible, or is it purely to generate maximum revenue? One may hope for both, but sometimes a product will receive a great deal of attention in the press, which increases visibility and awareness of the designer or company, yet which fails to generate great sales.

Big Shadow Lamp SE
Designed by Marcel Wanders. This famous lamp was produced for Cappellini in a limited edition of 150 pieces with a special gold-colored finish and silkscreened fabric.

One-off objects and limited editions connote good taste and denote an enviable lifestyle. In this way, limited editions or one-off designs can be promoted to create hype, and subsequently generate greater interest when the designer's next, maybe more commercial product is launched.

A good presentation of work can be the reason a product is given editorial space regardless of whether or not the product itself is the latest and best design. The design of magazines, papers, and books needs to catch the reader's eye, so great photography, styling, and attention to detail positions the designer or company in both the commissioning editor's and the end user's minds. Moooi, for instance, use a unique photographic style and well-thought-out promotional material to convey a uniform message. A carefully considered product mix that is both PR-savvy and has revenue-making potential will assist in engineering the career of the designer.

Giant Anglepoise
Designed by George Carwardine. To celebrate the 70th birthday of the original Anglepoise 1227, this large floor lamp, three times the size of the original design, was created. Produced in a limited edition of 250, each lamp is numbered and certified. With its effective marketing, this diversification raised the profile of the company.

Logistics

Logistics encompasses many supply issues: raw material supply and labor accessibility, the flow of materials or finished components within the production process, and the questions associated with timely fulfilment to a specific market.

As the costs of fuel and shipping increase and international supply and demand becomes more geographically dispersed, products need to be designed to reduce these inflating transit prices. How a product is packed and transported for delivery to both stores and end users is a relevant concern for the designer within the product development cycle.

Efficiencies in logistics mean that the end product can reach the customer quickly, at a more affordable price, and with less environmental harm. One approach would be to offer more licensed, strategic arrangements with far-off markets where the demand is great enough for this method of distribution. Production is closer to the final customer and shipping is kept to a minimum. This method requires a lot of investment, and issues with consistency in quality, both for the brand and for the product, need to be clear. This is more usual with a high-turnover product, such as fashion, that requires fewer specialized skills in production.

Another solution to this question of logistics in relation to proximity to the end user and the concern for design, is the maximization of logistics reach. Good logistics reach is achieved when a product has a high financial value and a low mass. Reduction of the overall product mass, including the packaging factor, is the most

efficient way to improve logistic reach. This area is particularly relevant where online sales are increasingly used as a channel to market.

So how can products be reduced in size? Solutions need to be found, as we are sure to experience a constant increase in the cost of moving product efficiently, with as minimum harm to the environment as possible. Designers must address these issues.

Flatpack as a concept has been adopted in the more affordable end of the interior furniture market with self-assembly products, but maybe this concept could be extended to other product types? Products assembled closer, or even within the end user's environment, maximize the cost savings in transportation. Designers need to embrace such means of improving distribution efficiency within the development of their designs.

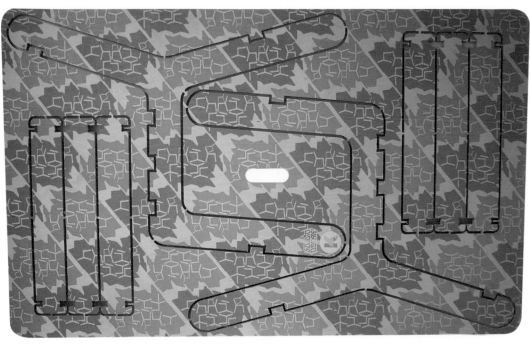

Left: Ply Bowl
Designed by James Harris
Designs. This bowl is formed
from two separate, lightweight
components that slot together.

Above and right: Chairfix
Designed and manufactured
by Ben Wilson. Three chairs
can be cut from one sheet
of plywood. The chairs are
reversible and offer the user
a choice of either a green or
pink repeating pattern. Such
choice adds value to this
piece through allowing greater
decorative options. Matched
with a design that incorporates
good logistic reach, and ease
of transport and assembly,
this collection transcends
perceived social and
geographic boundaries.

Distribution

It is important for the designer to consider the means of distribution when creating a product. The whole design chain should be considered holistically. Areas where design methods can be beneficial to the project as a whole can set apart the best designs and designers. Erwan and Ronan Bouroullec talked about the qualities of a designer as including a real capacity to work and see a project through from the beginning to the very end; to be able to engage in a good and informative dialog with all those within the development and supply of a product.

Information flow between the various areas within supply may offer a new insight or potential for development for a specific design. The Hanging Treillis for Teracrea, for example, considered observations on distribution channels and storage capabilities as part of the design.

Where a product is seen and how it is shown—from galleries to trade shows—is a vital consideration. Galleries can offer insights into responses to a design idea where the concept is more experimental—closer to art than product. Once a product is completed and ready for distribution to the end user, the companies involved in product visibility need to be identified. Producers take the goods either direct, through their own stores and distribution channels, or via the more complicated and usual process of supply, which might include the use of licensing agents, distributors, retail stores, galleries, and online stores. It is important for the designer to consider what happens to the product at these different stages.

Perhaps a product is made with a specific market in mind through a project with a client. This means that the product will, to a greater or lesser extent, have a predefined set of channels with that manufacturer, brand, or consultant. Established designers will have little concern with this process as often their products, if not created for a client, will find a buyer willing to take their work on due to the credibility of previous works. As a lesser-known designer, it is important to understand the whole process. What are the benefits for licencing a design as opposed to selling the whole design, or the differences between an agent and a distributor?

Distribution depends on many factors, often relating directly to the product itself, such as logistic reach or whether this is a one-off product or a whole family. Many questions need to be answered.

How should a designer present to a particular client and when is this best done? The general rule is to be well prepared and informative, remember that time and efficiency at all stages of the design process generally equal money saved or spent.

The diagram on page 136 outlines the basic differences and similarities within supply choices. This is not a concise listing but a starting point.

FLIT kite chair
Designed by Gudrun Lilja Gunnlaugsdottir. The lightweight FLIT can be folded and carried in a bag.

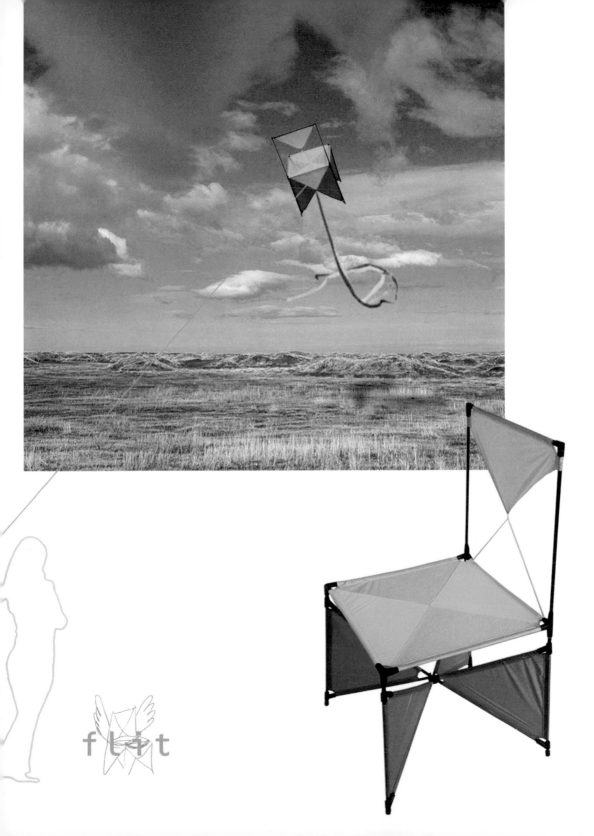

flit

	ADVANTAGES
AGENT	• Beneficial local knowledge of customers • Local knowledge of operations, including taxation and legislation • Established contacts for distribution and promotion • Payment to agent only on sales commissions
DISTRIBUTOR	• Localized logistics, product stock, and stationery • Improved economies of scale • Time to market reduced • Localized and established promotion routes • Payment in minimal batch quantities • Quick financial return on a design • Low market expansion and visibility costs
LICENSEE	• Financial and time savings as no manufacturer or distribution coordination required • Risks and costs spread
STRATEGIC ALLIANCES (shared production and distribution, reciprocal product and information exchange)	• Good exchange of information and expertise • Low cost development • Shared understanding

DISADVANTAGES

- High logistics costs, including stationery, information, and product flow
- Invoicing and credit control is in-house, not with the agent
- Increased costs in stock control and buffer stock levels
- Lack of relevance of agent product portfolio

- Possible brand image dilution
- Low control of distribution vehicles
- Increased production coordination costs
- Low control over sales prices
- Reduced product portfolio compatibility

- Low control over quality
- Frictions with partnerships
- Frictions over ownership

- Conflict of interests
- Difficult to assign control

Distribution chart
There are many channels to market for a product; this chart is an overview of some of the considerations.

Design rights

Intellectual property is a specialist area within design that requires research and legal advice. Companies invest heavily in their brands, and a product is the ultimate expression of the company image and therefore is an asset. But who owns the rights to the designs? How can designers protect themselves from copycats? How can they prevent and protect their designs from counterfeit? Healthy competition is needed for economic balance yet blatant copying is not.

In general, there are a few pointers for best practice although each country will have a particular legal stance and so international and export advice should be sought. Some protection can be found in copyright, patents, trademark, and design registrations, all of which vary in their degrees of success for legal defense. The first question to ask is that of cost and to balance this against the loss of a particular design right, because often to fight a case is more costly than the loss of the asset in the first place.

Patents: A patent holder has an exclusive right to apply a particular invention or even to prevent any other party from using it. The usual term is around 20 years, which gives the patent owner a long period to reap the rewards of this novel and useful invention.

Community Design Registration (EU only): This lasts for up to 25 years and can only offer protection when the appearance from a particular feature or ornamentation of a product is new and has individual character. This will only cover the members of the European Community.

Copyright: This does not need to be registered and is automatically given to work that is "fixed"—written, recorded, filmed, etc. Again, "fair use" is a term that is interpreted differently by country, as is the time period for copyright protection.

Trademark: This generally needs to be registered and should include a name or logo that has a degree of long-term use.

In a commercial world, where competition for design and innovation is fierce, the right to produce is bought and sold. This creates an ever-increasing need to identify ways of protecting designs so as to act as a real deterrent to counterfeiters. The visibility of products in broadcast, print, online media, and at exhibitions and shows means that using simple methods of protection makes good practice.

Copyright all photography, printed matter, and online material by page and image, register designs in the countries of origin and those that are intended for supply.

All protective methods cost money and time, but a designed product can be a lucrative asset. The question to ask is, how truly original is a design and is it worth the financial expenditure to protect?

Moomin mug
Manufactured by iittala Arabia Finland. The legal protection of a design should be clearly identified on a product.

Portfolios

The portfolios of work within the following section is a personal observation—a blink-of-an-eye snapshot of some of the most inspiring companies and designers.
The selection includes branded design companies, small-batch producers, manufacturers, makers, and designers. The companies vary greatly in size and international reach. The aim of this book is not to reveal who is selling the most pieces or earning the best returns, but to interpret what defines product design. It reveals companies and designers who embrace the idea of design as a dynamic and passionate narrative of self and surroundings. Some younger and less-experienced designers are showcased alongside well-known industry standards who have already achieved international acclaim.

Many of the design companies featured operate in Europe, drawing international design talent to certain cities, perhaps due to the concentrations of higher educational establishments. It is important to note, however, that the portfolios represent a truly international cross section.

The diversity of product type and the interesting approaches that are highlighted here illustrate the stages, the similarities, and the differences between designers and companies within the design process. The main thread that links all designers is that, within the goal to create something "new," there is a real need to be inventive. The term "new" in itself can be readdressed and questioned. To be inventive is a mind-set that not only embraces dynamic and free thought, but also acts upon it. Designers need to find different combinations of ideas and applications to have a good understanding of human sensibilities and to act as receivers for and broadcasters of the cultural implications of our time. The words "useful" and "beautiful" are also cornerstones of design aspirations, alongside design responsibility.

This section aims to provide a little design enlightenment—full of design eye candy with some captivating products, styling, and photography that all strive to answer the ultimate question: "What is Product Design?"

BarberOsgerby

Edward Barber and Jay Osgerby met while studying at the Royal College of Art in London in 1992 and set up BarberOsgerby in 1996. They formed a multidisciplinary practice with Jonathan Clarke in 2001, offering architectural, industrial, and interior design projects, and working with diverse and talented makers. Their definitive style incorporates a certain hand crafted lightness. Their product portfolio reveals a multitalented skill base, grounded in principles of innovative simplicity. It includes furniture design, rugs, decorative tiles, screens, laminate designs, bathroom accessories, and clothes hangers.

The team has a dedicated, holistic approach; they are involved with the whole process of creation, through development, manufacture, and launch. This contributes to a cohesively elegant style reflected in the total design package, which results in material forms that are machined and produced in numbers, yet have a handmade quality.

BarberOsgerby's first influential design—the Loop Coffee Table, originally produced by Cappellini in 1996—is a future design classic and now part of the Isokon Plus portfolio. Their contemporary designs can be found in permanent collections at the Victoria & Albert Museum, London, and the Museum of Modern Art, New York. Accolades include Best New Designer at the International Contemporary Furniture Fair, New York, 1998, and the Jerwood Applied Art Prize, in 2004.

Levi's commissioned BarberOsgerby to design promotional displays for the Levi's Red and Engineered denim ranges. Producing the Levi's Hangers presented a challenge: to determine a method of displaying the 3-D effect of this type of denim, which is a little harder to fold, and so altering usual merchandising and storage concerns. BarberOsgerby's solution was to create two injection-molded versions of a new hanger to display and store shirts and pants. Barber adds that the hangers needed to stack efficiently (200 to a box), as they had to be transported to over 9,000 shops around the world. In the development stage, between 30 and 40 molds and prototypes were discussed and tested before the final design was approved and made ready for production. BarberOsgerby rationalizes this concentration on prototypes. "We think you can always tweak a design for as long as you've got; the trick is to know when you've reached a point when it won't get better."

The Hula Stool reveals efficiencies in material use—its production involves little waste—and exemplifies the light and shapely form that characterizes their work; it applies a simple solution of cutting flat shapes from a sheet of ply. Compound curves give shape to the idea of a stool.

The commissions BarberOsgerby undertakes reveal their desire to take on fresh challenges and embrace diverse product types. Wall-fitted tiles, commissioned by Stella McCartney for a retail interior setting, show an adaptation of natural images of petal growth within a geometric, interlocking honeycomb pattern. Pattern repeats are a big step away from the curvature in furniture.

From this insight into the workings of Edward and Jay, there is much to be learnt about best practice and the development of a signature style across design media.

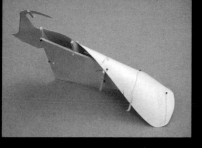

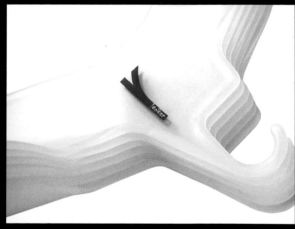

Levi's Hangers
Designed by BarberOsgerby
for Levi's. Shape-forming to
best show the 3-D effect of the
Engineered denim collection.

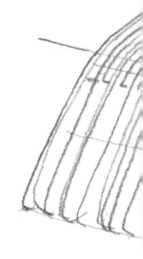

Below: Portsmouth bench
Designed for Isokon Plus.
In the past, artists used to be
commissioned to decorate the
interiors of places of worship;
now contemporary furniture
adorns St. Thomas' Cathedral
in Portsmouth, UK.

Right: Hula stool
This original sketch shows
how, sometimes, early
workings can be quite an
accurate representation of
the finished product. The
technical layout for CNC
cutting (center) shows how
sections are positioned as
tightly as possible to maximize
material usage in production,
and minimize waste. Photo
by Cappellini.

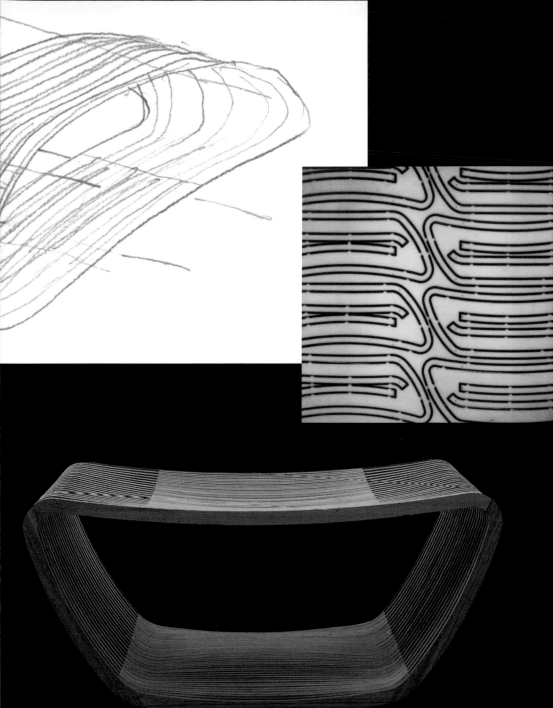

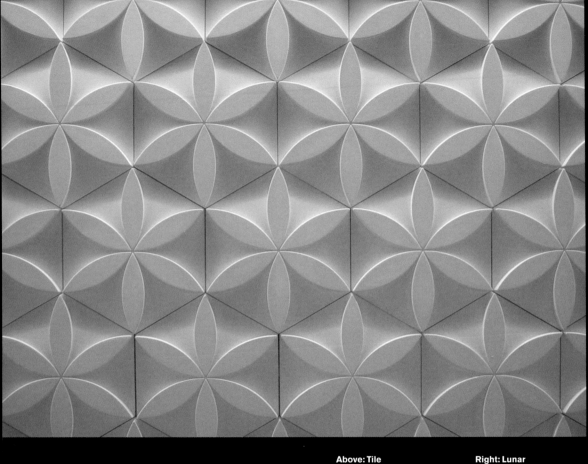

Above: Tile
Designed for Stella
McCartney. A bespoke tile
fitting that provides a signature
feature for the retail space.

Right: Lunar
Designed for Authentics,
Germany. Industrially
produced objects for everyday use. The green details
ensure that the product is less
sensitive to dust, and the white
endorses its hygienic function.
There is always a dialog
between the designers and
the manufacturer when
choosing the appropriate
finishes and colors.

StudioBility

StudioBility was established in 2005 to produce both product and graphic design. The team comprises a partnership of talents: Gudrun Lilja Gunnlaugsdottir and Jon Asgeir Hreinsson, a husband-and-wife collaboration. Asgeir Hreinsson works primarily within graphics-specific work and Gunnlaugsdottir specializes in 3-D design, heading up the majority of pieces in the furniture and experimental design portfolio.

"We try to be honest in our creation; it is the only way to be able to create something new," says Gunnlaugsdottir. This philosophy helps people connect with their work as they question the ideas of product heritage and the point at which a product becomes old and tired, and is thrown out of its orginal life.

StudioBility often experiments with ideas and speculative creations generated without a set brief. In such cases the production of readily available products might not be the ultimate goal, but rather, the production of inspiration and food for thought for other works. This process of experimental production without set parameters offers unparalleled freedom to produce playful designs that contrast with preconceived ideas of how certain products should take form. StudioBility challenges these perceptions, but always with attention to quality detailing. Its style crosses the boundaries of art and product design.

Gunnlaugsdottir explains. "What is extremely important for me is to play and let my mind go free. I start something and allow the inspiration along the way lead me. On later stages, it can be taken to 'down-to-earth' product design. Some I choose to take in that direction, some just remain playful."

The Camouflage Chair was an experiment with the ability to disguise and so fit into a certain environment—this principle was then applied to the Flower Chair.

Inspiration for much of StudioBility's work comes from nature and human behavior. The process of design begins at a very conceptual level and often the final design is surprising, breaking free from pre-conceived ideas. Materials used within the collection are applied with an eye to the full circle of production and waste. Inspiration for the choice of materials comes from considering the way in which materials can be utilized or given an extra dimension through the use of another process, as with the application of CNC decorative effects. Much inspiration comes from other media, but also from the "art of nature" or words. The main point is to listen for that extra beat, that which is missed without close observation. It is exactly that missed beat which is at the heart of StudioBility's creative process.

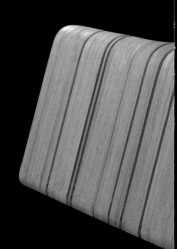

Left: Rocking Beauty
Rocking Beauty is an expression of body and soul; a piece in which femininity meets masculinity in form, decoration, and materials.

Above: Mermaid seats
The concept for these seats is the changing depth of water, and height of seat required, as the tide flows in or out.

Flat Pack Antiques
Surface decoration with
CNC machinery provides
technologically induced
individuality for mass
production. Ease of
distribution and assembly
is also addressed.

Vlaemsch()

Vlaemsch() is the design label of influential Flemish designer, Casimir, who has been producing product, interior, and architectural design under his own name, Casimir Design Office, since 1995. Casimir is also the design force behind upmarket hand-crafted furniture collection Casimir Meubelen, which was launched in 1995.

Casimir first showcased the Vlaemsch() range of products in 2004 at the Hasselt Showt Smaak in Belgium, with a collection from both national and international designers. The unusual use of greenhouses and catwalk presentations to display the new prototypes of this collection assured effective promotion for the brand. Attention to detail, with the production of informative print material for end users in a newspaper-style magazine, further propelled brand visibility to an international marketplace.

Vlaemsch() was borne of a philosophical approach to design, with a constant disguising and unveiling of similarities and differences between perceived values within our culture. A collection of affordable designs with mass appeal is the result of applying reductionist techniques that fragment ideas and concepts to the bare product essence, with a pinch of humor or a heavy dose of function: a birdcage that snaps together and takes form easily, a table that is stabilized by paperclips, a simple stool defined by the passage of time—all are observations of simple truths in altered states that beg to be noticed.

There is a definitive style to this brand that originates in the creator's own personality and values, but which is also fed by guest designers from Belgium, France, the Netherlands, Italy, and the USA. Associates include Fabio Baldelli, Frank Gielen, Max Borka, and Igor Foerier.

With his diversification into Vlaemsch(), Casimir has moved away from hand crafted objects and into a more industrialized production process, yet a voluntary simplicity underpins the collection, which always offers something interesting for the end user.

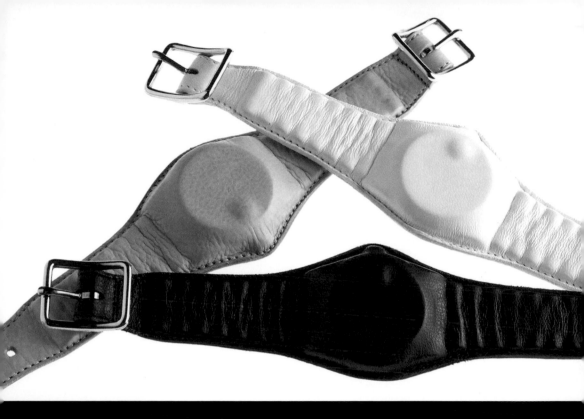

Above: Internal Rolex

Designed by Leon Ransmeier.
For his Internal watch,
Ransmeier encased a replica
Rolex in vacuum-formed
leather. Covering the watch
removes both its original
function and its typical
associations of wealth.

Right: Leather and plastic
monobloc chair

Designed by Front Design.
This plastic, monobloc chair
has a global presence; it is
comfortable, functional,
stackable, lightweight, and
hardwearing, yet its design is
unattributed. Front updates
and accessorizes the chair
with this leather jacket.

Annual stool
Designed by Leon Ransmeier.
Branded with a number, the
stool is identified and given
a greater relevance to the
question of "what is new

or old?" Here, the number 39
corresponds with the number
of annual rings from the tree
that gave life to the design.
This gives a small and almost
insignificant stool greater

potency and historical value.
A poetic expression of the
passage of time and the
value of the raw material.

Paperclip table

Designed by Casimir. Four
Trespa sections fit together
with 16 paperclips to create
a dining table. A reaction to
escalating shipping costs and
greater geological dispersion
of products throughout the
world, reducing distribution
expenditure by maximizing
flatpack potential.

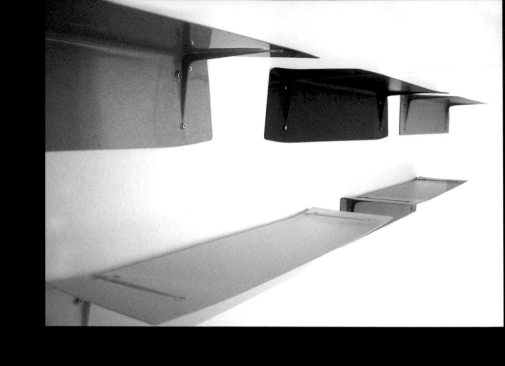

Left: Brackets Included
Designed by Sylvain Willenz.
A synthesis of shelf and
bracket. The inspiration came
from a scene in the movie
Barton Fink where the
increased room temperature
results in the wallpaper
peeling away from the wall.
Here the bracket and shelf
become merged and reformed
into one single unit, made
from powdercoated steel.

Below left: Birdhouse
Designed by Leon Ransmeier.
Two steel components fit
snugly together to create a
seasonal home for a bird.
A thermo-insulative ceramic
coating keeps the occupant
warm in winter and cool in
summer. This unusual
approach to marketing
endorses the Vlaemsch()
brand image.

**Right: Exhibition
installation**
Designed by Hasselt Showt
Smaak for Vlaemsch(). The
theme of the show, "living
in motion," encouraged
participants to consider
design and architecture
for flexible living.

Left: Fruit Bowl table
Designed by Casimir for
Casimir Meubelen. This piece,
a table in the form of a fruit
bowl, plays with ideas
of proportion and size.

Below: Trestle table
Designed by Casimir. An
economical table design
constructed from a door
and a set of trestles.

Dish n 3
Designed by Casimir. Using expanded polypropylene rather than more traditional materials, such as wood, keeps this large dish light.

Industrial Facility

Sam Hecht and Kim Colin established Industrial Facility (IF) in 2002 and have since been developing products and environments for clients, and working as creative directors for the European division of Muji, Japan.

Having studied history and architecture in LA, Colin moved to London as editor for visual arts publisher Phaidon Press. He also taught and studied industrial design at the Royal College of Art in London before returning to the USA, then moving back to London via Japan, working with Ideo.

In the 1990s the word "industrial" became associated with dirty factories and pollution, which led to its rebranding as "product design," yet Hecht still classifies himself as an industrial designer, at the disposal of industry.

Colin and Hecht interpret product design within the context of spatial considerations. They believe that the design of products should challenge formal product types and how they fit within a "product landscape," that is, how and where the product sits and interacts within a space. The IF approach is founded in observations of the everyday use and functionality of objects. Through research and identification of the softer, less tangible, even subversive nature of a product's components and users' interpretations of design characteristics, they offer something new. In their own words, "It is impossible to know whether an idea is successful until it is consumed. If it is possible, then the idea is not original."

The IF studio's portfolio of designs is created through identification of the essential characteristics of a particular project through the use of words, writings, materials, and models. Hecht comments that the usual starting points are "of this world," captured through photographs or drawings rather than dreams or fantasy. The design process follows a line of thought that is maintained and revised by in-depth research, analysis, experimentation, and dialog between everyone involved in the design process. Often, specialists are needed— engineers, model makers, graphic designers, business people—drawn from a pool of trusted, longtime contacts and collaborators. Experimentation, risk, and responsibility are adopted individually within the team.

The process is dynamic. Changes are often made up to and during the final stages of production, where even expensive tooling might be revised. It is a form of balancing and weighing scales of values to ensure that a project is the best it can be.

IF also works closely with clients, to the point whereby the client becomes a member of their design team. This is because IF consider design not to be a component of product development, but the very culture, the DNA of a company, and therefore an inclusive process.

Together, Hecht and Colin offer a stylistic narrative that goes far beyond product design. They often work on personal projects where the eventual goal is not for a particular client or production, but more a process to develop a library of ideas. Their ability to communicate through their design work reveals a peaceful and Zen-like appeal that can transcend industrial and product design to embrace wider, less-traveled design paths.

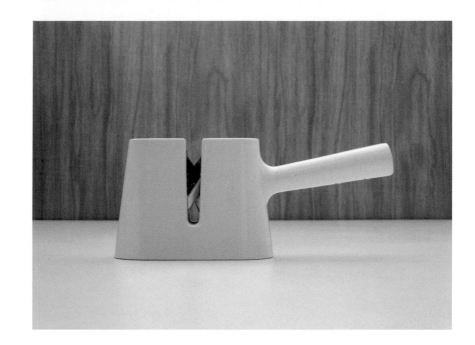

Left: Radius (PLL FM Radio)
Designed by Sam Hecht.
Concise, pen-sized radio that
adjusts its frequency to best
reception quality while the
user is mobile.

**Above: Chamtry Modern
knife sharpener**
Designed by Sam Hecht
for Harrison Fisher, UK. The
sharpening mechanism is
positioned within the design,
far away from the hand, and
the saucepan-like handle
provides a recognizable
and stable leverage.

Left: Industrial Facility
office London, 2006
Model laboratory.

Epson digital projector
Designed by Sam Hecht
for Epson, Japan, using ABS,
acrylic, and aluminum. The
hybridization of LCD and DVD
functionality means that this
compact, portable projector
can play your DVD selection
on any white wall where there
is a power source. The upright,
A4-size design resembles a
film projector with the spool-
like DVD player and the CD-
shaped remote control.
Familiarity is also gained
through the remote control
navigation, which is similar to
gaming controls. This design
exemplifies IF's philosophy of
"product as landscape."

IF400 Knife Program

Designed for Harrison Fisher, UK. From the models and drawings, we can understand the process of design through experimentation, analysis, and revision. IF integrated a fully forged stainless steel blade with a polyester and Melamine handle that gave a cool-to-touch quality. Research was in the balance, cutting angle, and ergonomics for this project.

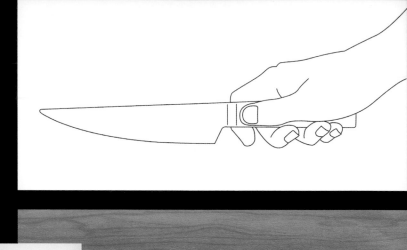

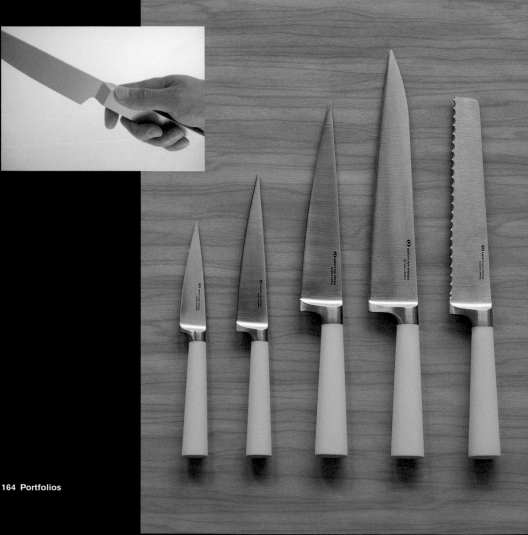

Key and Once Watch
Designed for Lexon, France.
Part of a collection of designs
of digital personal items. Press
the display itself for light.

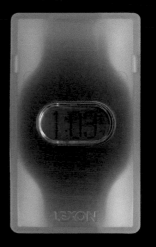

Moooi

The Dutch design agency Moooi is a young company with a multifaceted design strategy. Set up in 2001 by Marcel Wanders and Casper Vissers, the company is now owned 50:50 between the original directors and the influential Italian brand B&B Italia. The products are innovative and unique, and the company ethos is clear: they design the unexpected, have strong brand positioning, embrace the design process by welcoming many designers to create under the brand, utilize methods of color forecasting, and supply a holistically influential offering to an ever-growing international marketplace.

Moooi firmly believe that design is a reflection of our culture and a creator of the future. They understand that the key influences on design originate in culture, human need, and technology, and yet they also see how a potent message can be carefully engineered to create a unique stance within the design world. Products are reinforced by a vivid and inspirational promotional method—photography and collages push the boundaries of interpretation in their message. Moooi leads the way in effectively applying clever self promotion to get itself noticed and to create a pervading influence on its discerning customer base.

To look into the cultural mirror and reflect timely issues within customers' minds requires an ability to associate the past, the present, and the future. Moooi finds relevance in the application of existing objects, revitalizing and resurrecting them. Materials and new processes adopted from other industries offer the means of experimentation and innovation.

Moooi also offer a mentoring program to identify and work with talented people in the world of design, based on their philosophy that we are all both teachers and students.

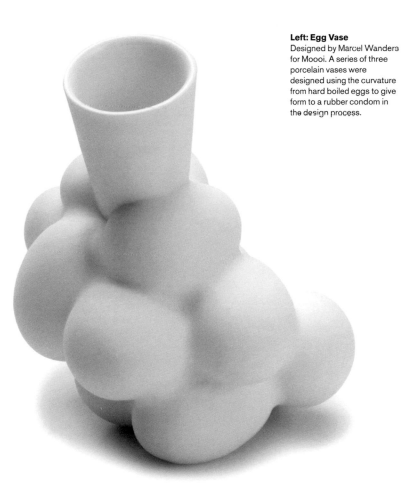

Left: Egg Vase
Designed by Marcel Wanders for Moooi. A series of three porcelain vases were designed using the curvature from hard boiled eggs to give form to a rubber condom in the design process.

Above: Bottoni Shelf Sofa
Designed by Marcel Wanders for Moooi. The seat has a curvature for greater comfort when sitting upright, and the arms offer a good headrest when lying down.

Left: Light Shade Shade
Designed by Jurgen Bey for
Moooi. The semi-transparent
mirrored film reveals a hidden
chandelier; when it is dark
and the light is on, the warm
glow and shape of a
chandelier is revealed.

Right: Flames
Designed by Chris Kabel for
Moooi. A candelabra with a
twist—with gas, the flame
never blows out. Designed
around a regular camping
gas cylinder. Photo by
Erwin Olaf.

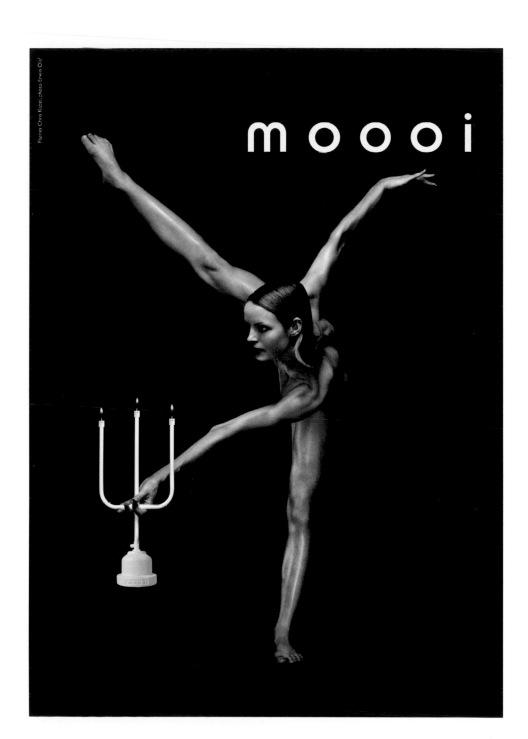

Flames Chris Kabel; photo Erwin Olaf

moooi

Above: Dear Ingo
Designed by Ron Gilad of Designfenzider for Moooi. A chandelier with extended functionality: the 16 lamps can be individually configured to provide a new meaning for directional lighting. Photo by Maaten Van Houten.

Right: Carbon Chair
Designed by Bertjan Pot and Marcel Wanders for Moooi. Made from carbon fiber and epoxy, this chair's light weight belies its strength. Photo by Maaten Van Houten.

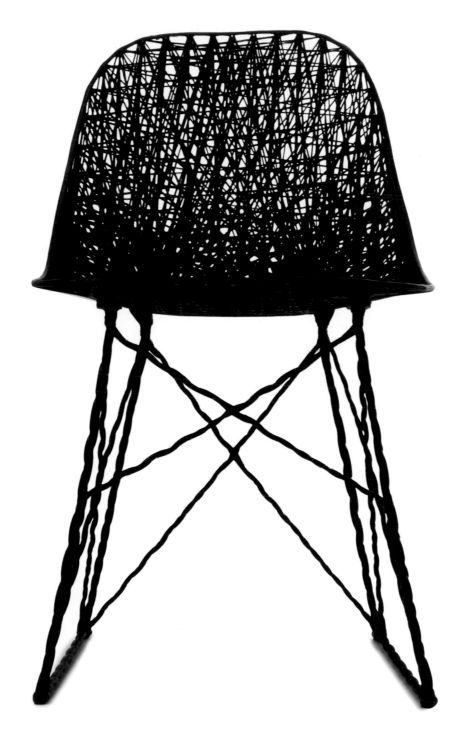

Smoke chair and Chandelier

Designed by Marten Baas for Moooi. Produced in an "alchemist" style, with fire consuming and smoking the wood frame, the result is never the same. The burnt wood is made functional by the addition of an epoxy finish. The black, silhouette-like furniture and lighting generate intrigue and mystery. Photo by Maaten Van Houten.

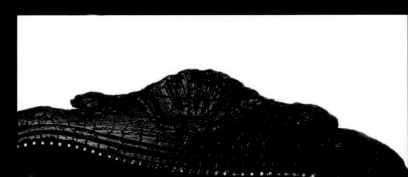

Biomega

Established in 1998 by Jens Martin Skibsted, Biomega challenged the traditional styling of bicycles, with a design-oriented, strategic approach to urban mobility; pioneered the values of minimum weight matched with maximum gears; and applied innovative technologies to the bicycle industry with night-glow frames, shaft transmissions, and dynamic, high-stress, load-bearing designs.

Skibsted believes that Biomega stands apart from other companies through its attention to quality, originality, relentless innovation, good design content, and the use of relevant technology. His approach is all about the quality rather than quantity of technology as a design driver. This is highlighted by the simple, technical functionality of his designs which offer something new without detracting from the form and function of the product. Biomega forges the way for the bicycle industry, not only as a technological leader and stylish bicycle producer, but also through its ability to make valid mobility observations within our cities and cultural society, and then apply these findings through symbiotic business collaborations.

This portfolio offers some insight into the design of the Puma Urban Mobility (UM) bike and its diverse product family. The family of products that are offered alongside the UM bike underpin the fact that this collection is a "segment buster" and offers practical functionality to many urban dwellers.

The UM bike came about through a mutually beneficial collaboration between Biomega, Puma, and Vexed Generation. Collaborations are magic when they work: all parties add to the overall value of the product at the end of the process, and offer the potential to cross over into their similar, yet different markets. The overall idea and brand concept originated within Puma. During concept phases, Vexed Generation brought in the "design against crime" angle, and Biomega designed the bike so that it incorporated Vexed Generation's forms and shapes while also adhering to Puma's brand heritage. Inspirations came from the messenger culture of urban life along with urban mobility, and the problems caused by crime issues such as bike theft.

Collaborative research and product development prove to be beneficial when individuals and companies work together on a particular project or on a regular basis. Puma and Biomega pared down the UM bike to its essentials and constructed an innovative new bike that applies rational design consciousness and matches this with street-cred esthetics. Unique features of the design include the locking system that forms part of the structural framework of the bike, an upright navigational posture for the user, a folding mechanism for size reduction, bike accessories, and apparel for the user.

The future for Biomega may well involve more unique collaborations that uncover real, value-adding observations about society.

UM bike by Puma
Designed by Jens Martin Skibsted. The new locking system safeguards the bike from theft and simplifies mobility as it eliminates the need for a separate, and often heavy lock. The lock is now an integral part of the bike! An easy-to-use, pin-release action frees the locking wire, enabling it to be wrapped around a stationary object and secured back into the bike.

UM bike by Puma

Designed by Jens Martin Skibsted. Biomega, Puma, and Vexed Generation teamed up to produce the UM (Urban Mobility) bike. The sleek, minimalist bike is lightweight and can be folded, resulting in a 50% reduction in size to fit in tight spaces—great for commuting and storage.

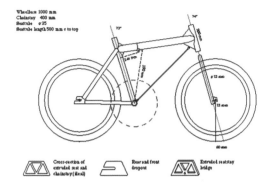

Biofold 24" ertro 507

Wheelbase 1000 mm
Chainstay 400 mm
Seattube ø 35
Seattube length 500 mm c to top

Cross-section of extruded seat and chainstay (ideal)

Rear and front dropout

Extruded seatstay bridge

Folded frame

Biomega carrier
Designed by Jens Martin
Skibsted. This luggage rack,
designed to fit the Urban
Mobility backpack, is
lightweight, and easy to
attach. All the Biomega bike
accessories are designed to
integrate perfectly with the
Biomega bikes.

**Urban Mobility backpack
and Lumbar bag**
Designed by Vexed
Generation as part of the
collaboration on the Puma
UM bike project.

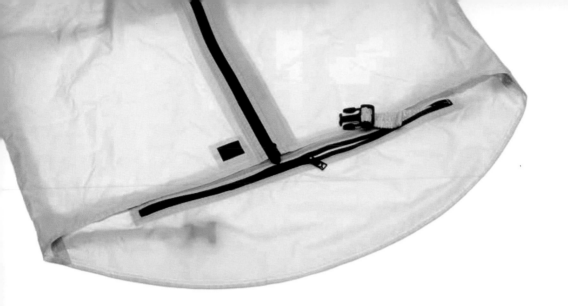

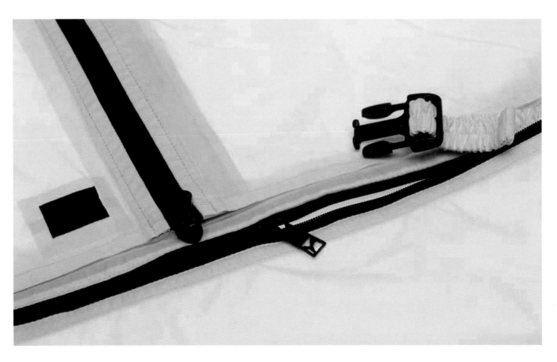

Pack-away jacket
Designed by Vexed
Generation. These jackets
reflect the emerging crossover
market between design, sport,
and fashion.

Front was established in 2003 by Anna Lingren, Katja Sävström, Sofia Lagerkvist and Charlotte von der Lancken. The four met while studying industrial design at Konstfack School of Arts in Sweden. With each member having come into product design from a different discipline, Front has a diverse knowledge base that it blends to create a distinctive design style.

They have produced collections that reveal a more unusual approach to idea and concept generation, involving the incorporation of nature within the design process. Of course, nature plays a role in every aspect of our lives and is an integral part of all design, but it is rarely embraced in such an unusual way—defining visual expression through the interactivity of natural processes. The first collection, Design By, marks the starting point of Front's work—questioning the idea of the designer as creator. Through research and the adoption of random influencing factors, Front has developed a diverse range of products—lighting, furniture, audio equipment, and decorative objects.

This application of nature through design is part of Front's design philosophy, which is to communicate a story about a particular process, conventional product, or material. Front also applies techniques and materials in unique ways.

The team explain their driving force in this way: "We are curious about the functions of an object other than the strictly practical. Why do you choose one thing over another with a similar function? What makes you care about an object?"

Front's ambition is to add emotional value to its products. The designers' approach follows the principle of uncovering something "new" through discussing, within the collective, topics that hold a personal interest for the members. They collate information and facts to work up an idea or concept. "We always work with different processes to develop our products. We usually work with things that we have little knowledge about."

Front has a permanent installation at the Röhska Museum in Gothenburg and exhibits around the world through trade shows and galleries.

Top right: Story of Things
What defines a home?
"We wanted to look at the
similarities and differences
between items from homes all
over the world. What happens
with products after they have
left the store? Objects tell
stories of moments in our lives,
people we have met, and
places we have been to." The
objects were reproduced in
red plastic and the relative
story printed. Front collected
stories over a long period of
time that explore the age-old
dichotomy of love and hate.

Left: Design by Gravity
A lamp that reacts to the
user's movement—when
inactive, the lamp lays down,
when awake, it lights the room.

**Below right:
Enlarged cushion**
A greatly oversized cushion
becomes a seat for two.
Photo by Anna Lönnerstam.

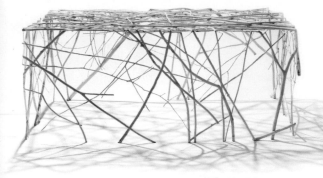

**Above and right:
Design by Pressure**
A new material made from
branches by using high
pressure. The material can be
pressed in a mold. Here it is
formed into a coffee table that
casts shadows like a tree.

Left: Design by Mechanics
A table that acts unexpectedly
and does things on its own
terms: at first, it stumbles
and after a few days it learns
how to walk.

Insect Table
The insect's passage through the timber creates a decorative pattern. Photo by Katja K.

Design by Surroundings
The decoration on the vase is made of permanent reflections. The vase tells us some of its history, the new reflection blending with the old like a double exposure.

Design by Sunlight
A UV-sensitive wallpaper
where the pattern changes
with the sunlight. The pattern
made of shadows from

Ronan & Erwan Bouroullec

Ronan, the older of the two Bouroullec brothers, started his design career having completed a course in industrial design at the Ecole Nationale Supérieure des Arts Décoratifs, in Paris. After completing a course in furniture design at the Ecole des Beaux-Arts, Paris, Erwan joined Ronan as an assistant in 1998. By 1999, they were creating designs together. Most of their work is interiors-based products and interior design. Their diffuse collection of products over the past eight years includes: lighting, desks, tables, lounge chairs, jewelry, carpets, and accessories.

The speed of the brothers' success is due in part to the early adoption of their ideas and creations by Giulio Cappellini, who first met Ronan at the Salone Satellite show in Milan in 1997. Companies that have produced their wonderfully empirical and stylish designs include: Magis, Habitat, Ligne Roset, Issey Miyake, Vitra, and the Kreo Gallery in Paris.

The Bouroullecs' design process is based on maintaining a dialog with all of the parties concerned in the creation of the final product, and most importantly, on their ability to work together productively with each dynamically challenging and complementing the other. The dichotomy of similarity and difference between them, especially with disagreements concerning a specific project, is ultimately highly productive: it leads to intense discourse in finding the best solution for a particular problem. By the nature of this inherent dialog, every detail and principal behind each facet of the design process is well versed and has been developed to the point that presentations with clients become natural.

"Our work is nourished by our constant dialog," confirm the Bouroullecs. This strong and skillful dialog means that, while working with specialists and manufacturers, a real and in-depth knowledge is generated to feed the design process. Understanding the considerations of the client, manufacturer, or end user (whoever is involved in the collaboration), along with those of the process and materials used, adds to the end value of the product ,and often a certain characteristic is uncovered that might define an object. "We try to observe people's behaviors in everyday life in detail and to understand usual practices and needs. Observation is our main external influence."

Sketching is a tool that translates their ideas into detailed realities, with numerous drawings for each design; a notebook can hold a multitude of design ideas at various stages of development, where features can be identified and clarified.

The Bouroullec's tools for shaping their ideas are "Two brains, a pen, some paper, a computer, and the usual modeling tools."

Through intense thought, sketching, and dialog, concepts are developed. Products are defined by questions, and solutions are uncovered through the design process. Different product strategies and features result in pure and fluid creations that are engineered for durability of design and function.

Cloud modules
An abstract repeating
structure is created from
individual modules that
are inspired by repeating
forms in nature.

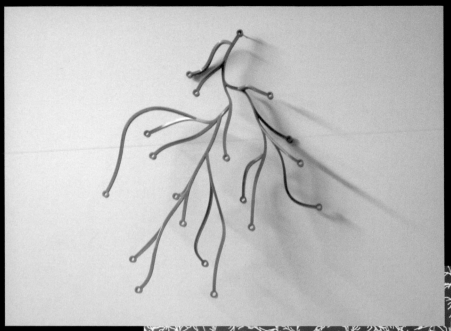

Algues
Injection-molded units that can be assembled to provide a decorative screen of any size. Repeating plant life that references sub-aquatic space.

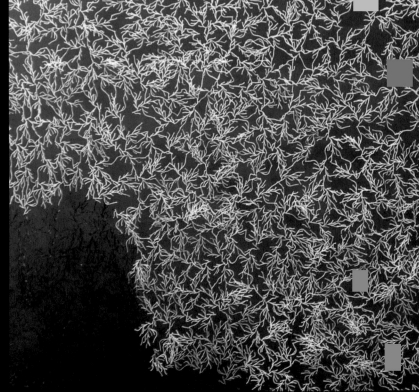

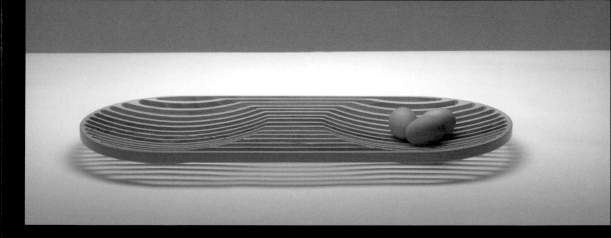

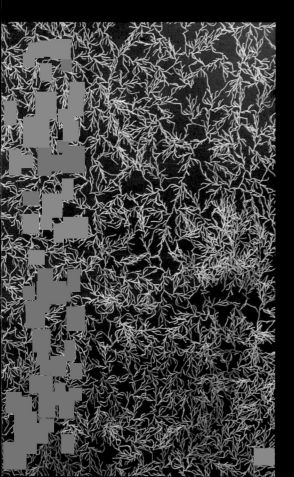

Fruit Bowl
Designed for Cappellini.
This bowl rests on its slats,
and the groceries placed
on them can breathe.

Grape Carpet
Designed for the Kreo Gallery, Paris. The Bouroullecs are interested in ease of assembly because it is fundamental in terms of the logistics of manufacturing and distribution, and because it gives the user a greater degree of choice.

Bells Assemblage
Designed for the Kreo Gallery,
Paris. Hand-embossed copper
is lacquered in aubergine, dark
blue, or light blue into lamps
and corresponding side tables.

Above: Hole Console
Designed for Cappellini.
This wall-fitted shelving
system, part of the Hole
collection, shows a more
traditional approach from
Ronan & Erwan Bouroullec.

Miller Studio

Jason Miller set up his design studio in New York in 2001. His work references everyday objects that are given new meaning and visual context by their surroundings.

Miller designs within the realms of conceptual art and explores observations pertaining to the way we view, use, and produce objects. This approach reveals the play between the reality of functionality and customization of a diverse collection of objects—toasters, teapots, tables, lounge furniture, mirrors, antlers, etc. Making use of the crossing and blurring of creative boundaries, innovative products arise from different skill bases, including art, graphic design, and crafts. The collection of works here shows the play within these boundaries.

Miller usually works alone using a hands-on design process, starting with the uncovering of a general concept. Deep thought and annotated lists of ideas, thoughts, or even products are accumulated until something interesting pops out of the page. At this point, the idea is clarified, sketched, and sometimes modeled with a 3-D program to arrive at a visual identity. "Usually my 'sketches' are more like a collage then a drawing," he says. "My favorite tools are my Computer (a Mac G3 most of the time), my camera (Canon Power Shot), and Duct Tape."

As well as designing products, Miller also curates shows that give American designers greater visibility, including Living Spaces within the International Contemporary Furntiure Fair (ICFF). The Miller Studio collections are often exhibited at solo shows within contemporary galleries, and this has helped build up a conceptual, arts-based image for his work.

"It's not enough to just be a good engineer or a good woodworker or a good shaper," he says. "Product designers have to deal with a broad range of problems when making a product and therefore have to have a broad range of skills."

**I Was Here Table and
Reference Graffiti**
Inscribed and embellished
park benches, bars, and table
tops from New York City
provided the inspiration and
the graffiti for this design.
Recycled plastic with
CNC-routed graffiti offers
symbolic homage to past
moments in time.

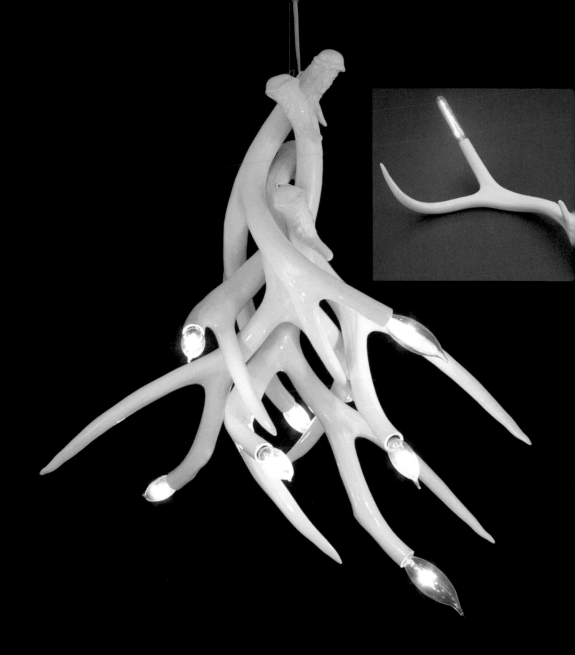

**Superordinate
Antler chandelier**
Produced from antler molds
to create a natural ceramic
light. Originally designed as
part of a show that explored
the relationship between
urban and rural design.
Available in six- or four-antler
combinations, table and
sconce options.

Seconds

A series of teapots and tea sets decorated in the "wrong" way. Decals are cut up and applied over the surface to show half a bird or a plant that seems to grow toward the floor.

Above: Mismatch Chair
Miller's notes for the
upholsterer illustrate how
to achieve mismatched
pattern effects: "It shouldn't
take seven yards of fabric to
cover this chair. It's not that

Right: Lego Vases
Porcelain vases cast from
Lego sculptures. "Lego was
one of my favorite toys. I could
make anything, huge cities or
tiny cars, creatures, boats,
whatever I wanted," explains

Muji

The Japanese store name Mujirushi Ryohin translates literally as "no label, quality goods." Known as Muji internationally, this is a company with a strong Japanese cultural philosophy that translates through the whole brand package. Muji is synonymous with a product range that encompasses concepts of standardization, ingenuity, good value, and timeless quality.

The timeless qualities are engendered through the conscious understatement of color, and the development of absolute quality and functionality along with a "what you see is what you get" marketing style. There are no hidden meanings and no layering of effects in the products, packaging, advertising, or store layouts. Such a cohesive company image reinforces the customer experience of reliability and problem solving for certain product types.

Efficiency in production processes and the application of industrial materials, matched with an effective design process, mean that creative and original products add real value. One of the better-known designs from Muji— the wall-mounted CD player—illustrates the company philosophy. This product won the iF Product Design Award in 2002 and was relaunched in 2005 with a limited-edition black version as a form of diversification. The pull-cord interaction is obvious to the user as the player borrows its principles of operation from ceiling fans and light cords. The design plays with the visual and auditory effects of product interaction.

The original idea for the speaker came from the observation that people store many hundreds of CDs. The main focus of the project was to produce a speaker with dimensions in direct proportion to a CD case. Muji has developed a design process that exemplifies dynamic and free methods of working which build on the need to design the unexpected. Product design consultants Seymour Powell assert that we need to develop these new, different ways of practice so that new solutions can be entertained.

The Muji method is "without thought," yet the radio CD player is an imaginative product that combines pared-down simplicity with added visual functionality—the speakers also act as book ends. The concept embraces the development of empirical forms that sit quietly and discreetly in any home while adding extra value to create product desirability.

Muji's star is on the rise with globalization of the company through stand-alone retail stores, concessions, and partnerships, along with a strong and consistent identity.

Wall-mounted CD player
This compact and functional design has no superfluous elements. The speakers are built into the body of the CD player and to make a disc play you simply pull the cable.

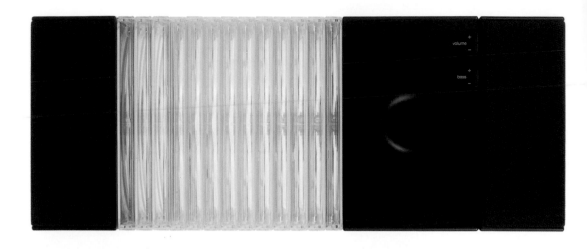

Above: Speaker set
These simple, basic speakers
have the same height and
depth as a CD case so make
perfect "bookends" for your
CD collection.

Left: CD and Radio
Simple, uncomplicated
shapes are characteristic
of Muji's designs.

Washboard
This understated washboard
continues Muji's tradition of
sleek, functional design.

Waterproof speakers
The showerproof polypropylene encasing the speakers also forms a protective case into which you can plug a CD player.

Paper shredder
This simple paper shredder
forms part of Muji's
comprehensive range of
coordinated home office
and stationery products.

Royal Tichelaar Makkum

Founded in 1594 to produce bricks, this is the oldest company in the Netherlands. Toward the end of the 17th century, the factory was adapted to become a pottery and has since produced ceramics. Royal Tichelaar Makkum's (RTM's) range of services and products includes restoration of historic designs, bespoke architectural and product commissions, and a diverse collection of both classic and contemporary products, including decorative ornaments, tiles, and functional vessels.

A company is not static; it is a living organism that needs to respond and adapt to an environment. Traditional, well-practiced, and skillful techniques, along with the commissioning of some of the best-known Dutch designers, allows RTM to update its product portfolio stylistically on a regular basis. Matched with the company's attention to contemporary work is the establishment of continuity through its continued production of design classics, which move in and out of fashion over time. In addition, the skills involved in the production of its contemporary range are the same as those used throughout the company's product history. Influential Dutch designers—among them, Hella Jongerius, Marcel Wanders, and Jurgen Bey—are commissioned regularly to add to the vast collection of products. They work with techniques such as majolica and Delftware, utilizing crafted details that are often hand-finished using skills handed down through the generations.

At RTM, Delftware following the traditions of the 11th century is created by shaping and firing Frysian marine clay, glazing this with tin, hand-painting the required decoration, and then applying another, final firing. Ancient traditions will always inspire designers; RTM successfully adopts age-old production methods and craftsmanship for the creation of fresh, contemporary designs.

Red/White Vase and Big White Pot
Designed by Hella Jongerius. Part of the process is revealed with the mold imprint left on the vases. Dramatic use of color and the stamp of authenticity give the pieces greater presence.

Soupset
Designed by Hella Jongerius. Nine matching soup cups and a ladle. Jongerius based her designs on authentic majolica techniques that reference domestic ceramics from the 17th and 18th centuries. Partially immersed in tin glaze and hand-painted, some pieces are more or less painted than others. The sailing ship motif was found in the company's library of old décor books.

T-set
Designed by Hella Jongerius. Made from porcelain and aluminum. The T-set continues Jongerius' experimentation with creating a handmade feel for mass-produced products.

Time series (65, 890, and 1,132 minutes)
Designed by Jurgen Bey.
Hand-crafted work takes
time to produce. Bey created
a hand-painted tea service
collection identified by the
time taken to produce each
piece, utilizing molds, relief-
printed techniques, and the
skills of the in-house artists
to paint each piece.

**Left: Sconce set
(screenprinted, hand-
painted, and white)**
Designed by Job Smeets.
A historical piece becomes
rather decadent with the
printed jewel accessory. The
hand-painted work still looks
wet from the artist's brush.

Bottom left: MaMa vase
Designed by Roderick Vos.
A potent use of color adds to
this friendly and ornamental
vessel. Its shape was inspired
by organic forms.

Below: Piggy Bank
Designed by Job Smeets.
Part of the in-house collection
from the series Still Life.
Piggy Bank is also produced
in screenprinted and hand-
painted versions.

Flo

Established in 2001 as a consultancy for new designers, Flo has since become a brand that represents, sells, and produces products ranging from wallpapers and textiles to accessories, furniture, and lighting. Inspiration for the business came from the realization that, despite the UK having some of the finest further education institutions in the design world, very few British designers actually continued their design careers after completing their education. Flo discovered that many simply could not find work, and others could not afford own-brand production due to high set-up costs. Flo also found that there were few production companies willing to take the financial risk of manufacturing and distributing pieces by lesser-known designers.

One of Flo's first projects, working with Sebastian Wrong on the Spun Lamp, established it as a company that engineers and builds on design talent to position and develop a successful collection. Following its creation, the Spun range was sold to Flo.

Flo has since expanded its production and distribution portfolio, and has a growing international following and market penetration. An unusual company, it understands and embraces all aspects of supply and demand within the industry. It produces and distributes its own brand, to both contract and retail clients, and acts as agent, manufacturer, and distributor for other designers.

The company also has a retail presence, Places and Spaces, in London, established in 2003. The store enables the company to build rapport with end users, and to research and develop products that fit their marketplace. It tests prototypes and uncovers demand for product niches that, historically, have had few choice offerings. An example of this is its diversification into contemporary paper hangings. Flo anticipated the trend for pattern revival in 2003 with the launch of the Seasons wallpapers, by graphic design company absolute zero°. The initial idea behind the project came from the need to break down the boundaries and pigeonholes within the design industry. A brief was sent to selected graphic, textile, and product designers asking them to participate in creating a new style for pattern with the potential for end use on wallpapers, textiles, and accessories—usually the skill of a textile designer. The wallpaper range has since been developed into a wider portfolio that embraces traditional techniques of paper printing, including surface print and flocking methods, with contemporary patterns.

Responding to market demand, Flo also commissioned Paul Croft to design a desk that was affordable, and that fulfilled predetermined criteria uncovered by market research undertaken at the store. The end result is a functional, honest shape suitable for both domestic and contract end use.

Some of the international guest designers working with Flo include twocreate, Alex MacDonald, absolute zero°, Donna Wilson, Tomako Yamashita, Mark Cox, Stephen Cheetham, and Paul Croft. The company plans to expand their own-brand products and gain a greater international customer base while also developing the retail brand Places and Spaces.

Left: Plant Pot Umbrella Stands

Designed by Stephen Cheetham. Flowers are watered by run-off from an umbrella. The steel frame also acts as a support for the growing plants while offering umbrella storage. The stands and pots can be purchased separately, so this design can be easily packed and shipped with low cost.

Right: Proud

Designed by Mark Cox. Standing proud, this floor-standing light allows the user to display an object they consider precious—an expression of taste and personality. Hand-turned in FSC-sourced timber, there are walnut, and white- or black-painted options.

Seasonal wallpaper series
Designed by absolute zero°.
The three designs are all
formed by repeated motifs.
In Swallows (left and detail)
falling leaves morph into birds,
then swallows; in Bees
(center) a honeycomb
structure morphs into bees
and butterflies. In Dandelions
(right) a dandelion clock motif
explodes into dandelions.

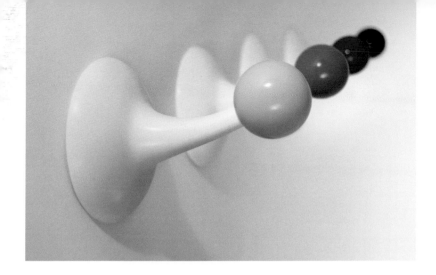

Peep hooks
Designed by Tomako Yamashita. A friendly shape cast in Corian with colored detail. Photo by Ian Rippington.

Doyley Rug
Designed by Donna Wilson. Having graduated from the RCA, Wilson worked with Flo to find a method suitable to produce this 100% wool-felt rug with detailed cut-away sections to offer contrasting color and texture to flooring. This design works with the user's choice of wood, resin, concrete, or tiles by complementing rather than covering up.

Left: Scandia Senior chair
Designed by Hans Brattrud. Curved, laminated wood slats conform to the user's shape. A black leather cushion adds to the comfort.

Azumi

Tomoko and Shin Azumi first met in 1989 at the Kyoto University of Art, Japan, where Tomoko studied environmental design and Shin product design. They then moved to London to study furniture and industrial design at the Royal College of Art, and set up design consultancy Azumi in 1995.

Having studied different specialist areas of design, they complemented each other and created innovative designs within many product typologies. Their work is retailed internationally, and is also on show at galleries and museums including Stedelijk Museum, the Netherlands; the Victoria & Albert Museum, and Crafts Council Gallery, London; and Die Neue Sammlung, Germany. As well as work created for manufacturers and producers, Azumi have also produced a few limited-edition, short-run production pieces independently.

Their designs originate directly from commissioned briefs for specific product requirements, or through their own observations of everyday life. Often, the designers define the brief for the client, developing dialog and clear communication from the beginning of the process to the benefit of the end design. Their work is innovative, with subtle attention to detail that derives from their observation of the way users interact with a product, and that quietly promotes a less obvious product feature.

Minimum styling gives their designs a certain honesty. The team consider product longevity as well as maximum usability, and take time to uncover all the phenomena around a project in order to clarify concepts for development. This period of design is not just about the actual product makeup, but also situational influences. Through the distillation of research findings, the duo are able to prioritize and engineer the concept behind any given design. They then deliver paper maquettes of shapes and design principles to their clients to explain and give visual identity to the concepts within their work. As can be seen from the models shown here, such detailed 3-D visualizations provide a vehicle for communication for the sculptural esthetics of the designs, and also show the design's real functional context.

Ten years on from the establishment of Azumi, and with a highly considered portfolio of designs to their name, the pair decided to set up separate studios at the end of 2005. Shin Azumi's "a studio" works across product, furniture, and space design, and Tomoko Azumi's "t.n.a. design studio" works across product and furniture design.

Shipshape and Penguin Donkey 3

Designed by Shin and Tomoko Azumi for Isokon Plus, UK. Shipshape (left) stores magazines and newspapers, and conceals remote controls, pens, bottles, etc. The structure provides enough strength for Shipshape to double up as a seat or a step. Only 100 of the original Isokon Penguin Donkeys (below left) were produced (in 1939, by Egon Riss and Isokon founder Jack Pritchard). Mark 2 was designed by Ernest Race in 1963. Pritchard gave Azumi a few indications as to the direction for the newest redesign: a more modern, easier-to-produce, flat-top side table. Mark 2 was designed with an associated product, the Bottleship. When Isokon Plus commissioned Azumi to evolve the Penguin Donkey, they also redesigned the Bottleship, resulting in the Shipshape. Azumi needed to incorporate in-depth knowledge of the historical perspective of the original work within their design. Approximately 30 paper models (below)—products in their own right—were created to reveal the thought process behind their designs. A true artisan's approach.

Above: Maruni chair
Designed by Shin and Tomoko Azumi. Embracing technological advances in production with 3-D wood curving and CNC routing, this challenges our stereotypical perception of a chair. Upholstered seat and back with wood frame.

Left: Trace chair
Designed by Shin Azumi for Desalto, Italy. As Shin Azumi explains, the form is like the trace of the outline of an armchair, hence the name.

Megaphone

Designed by Shin Azumi for
TOA, Japan. A megaphone
with the ease of use of a
microphone. The clear horn
gives greater visibility for the
user, and it also has a wall

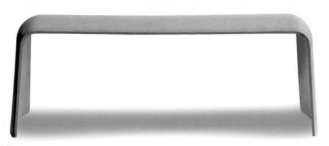

Above: Volta bench
Designed by Shin and Tomoko Azumi for Desalto, Italy. The Volta is formed from a single piece of ply.

Right: Yauatcha Tea and Sake set
Designed by Shin Azumi for Hakkasan restaurant, UK. Commissioned to complement the sophisticated design of Chinese restaurant Hakkasan, these pieces display Azumi's signature minimalist style.

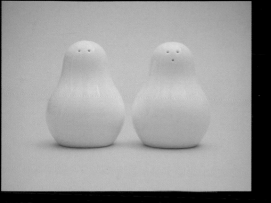

Snowman series
Designed by Shin and Tomoko
Azumi for Authentics,
Germany. Refreshingly original
salt- and pepper shakers.
By moving the holes further
down the vessel, an emotional
context is given to the objects
which now resemble snowmen.

Youmeus Design

Established in 2000 as Future Creative, the company was rebranded in 2004 as Youmeus Design. As the name indicates, Youmeus is a human-centric design company that creates products through an understanding of people and their needs, wants, and feelings. Developing products that cross specific design boundaries—from homeware to consumer electronics to transportation vehicles—Youmeus Design works with a total design philosophy that transcends traditional working practices. Creating products for both small and large companies, the main emphasis of their work is to make a difference, to challenge, and to question what constitutes good design. In their words, "People, the consumers and users, need more than just solutions; they need to be wooed on an emotional level. If a consumer is moved emotionally by a product, they will invent reasons why they must have it."

Working almost with the principles of a collective, the design team is supported by carefully selected business specialists, through to engineers for each project. This method of collaborative, inclusive design production brings into question the definition of a designer: is a designer purely a maker or a visualizing conduit, or is the concept of design now integrated as a strategic philosophy throughout a company?

Youmeus Design maximize their ability to work harmoniously with clients by outsourcing specialists and working as a team to achieve consumer-centric products. Effective teamwork adds value to all areas of the design process—with effective communication comes knowledge sharing, which allows ideas to germinate and reach full potential as both process and product are continually refined from many perspectives. "The job of the designer is to help identify the motivations of the consumer and then, more importantly, to add the x-factor to the product solution."

Youmeus Design adopts a holistic design approach and creates this wow factor with mindful exploration through research, visualization, assessment, definition, and delivery of a product for a specific project. The application of knowledge gained from research offers consumer and production insights that can uncover x-factors. Youmeus Design builds relationships between people and technology to make intuitive products. The integrity of their designs are, therefore, consistent throughout the process, and Youmeus Design are able to offer additional design advice to their clients in the final stages of production. Their products offer rich experiences with interactivity and usability, built through a thorough and up-to-date knowledge of trends, behaviors, attitudes, and technologies.

Chef food mixer
Designed for Kenwood. This design evolved from a classic product while maintaining its versatility and brand identity.

Travel kettle

Designed for Kenwood.
A discreet travel kettle with
carry case–the ultimate
space-saving design. The
kettle is part of a range of
travel products that challenge
the more traditional miniature
travel versions. By pairing the
kettle down to just the heating
element, its typology is
made unrecognizable.

Hand-held, powered kitchen tools

Designed for Kenwood. The whisk, electronic can opener, and electronic knife, are made from ABS, rubber, polypropylene, and steel. These kitchen tools adopt a prosthetic look—the tools become an extension of the user's arm and hand.

Portia Wells

Born in 1980 in southwest Washington State, and now living and working in Brooklyn, New York, Wells creates domestic pieces that reflect her views on class wars. Her work explores the way we interact and communicate within society and how we view certain products as cultural identifiers that distinguish and unify us.

"My work is inspired by what we consume and the promise of popular culture and labels," she explains. Wells uses a conceptual ground to originate her work and applies a process of making that she explains as a sort of affliction, where nervous energy becomes focused. As each project requires a unique outcome, she doesn't use any set pattern or tool for planning, but rather an observation of dedicated silence and thought. Researching theories, histories, and existing products results in an object that embodies uncovered issues that are important in an appropriate form.

When asked about the external influences on her work, Wells is exocentric, with a need to identify and understand her universal surroundings and mirror her findings through product design. Key influences are space, time, place, money, need, and fulfilment.

Some of her work embraces issues of sustainability, giving found objects a new and functional twist. Other pieces employ customization to alter a perceived value, or even question the difference between architecture and interior or product design. Questions are provoked by the design process. Wells stresses the importance of constantly reviewing the social norm in order to nurture and alter the value of a product.

Once Wells finishes a particular work, she usually covers it with a blanket or hides it away for about one week. Time allows for the mind to take on a slightly different perspective and notice any necessary alterations and final touches.

Wells' collection is sold through contemporary boutiques and Web sites throughout the USA.

Pen and Pencil Holder
There are many vessels
throughout our homes that are
already perfect for the task of
holding pens and pencils. This
brought into question the need
for a new form. The answer
was to differentiate the object
from others, to label, signify,
and reinforce the use of the
object. Found vessels are
given a breath of new life with
gold decal labeling.

Left: Slipcover Chair

The printed slipcover offers the user the ability to change the façade of a chair, whether it's a Chippendale, an Eames, or a Shaker, to a simple metal folding chair.

Above: Portraits

Where we live, what we buy, and what we wear act as labels, conspicuous or not. Wells notes how these items define our self-portraits. The use of silhouettes intrigues and draws the viewer in to

Left: Wall Lamp
Wallpaper by Wook Kim.
Graphic, textile, and fashion
designers all create patterns,
but what about architects,
interior designers, furniture
designers, and lighting
designers? And what
about the possibilities for
integration? The wall light is
not a separate entity, but is
wallpaper, wall, and light.
Photo by Glen Jackson Taylor.

Above: Inkblot
A classic damask ground
camouflages 10 flocked
inkblot patterns, inspired
by the Rorschach inkblots,
a popular psychiatry tool.

Re Design

With the increasingly positive perception of sociopolitical movements, remarkable words are coined to identify areas for potential design change—reclaim, recreate, recycle, reduce, remake, remind, resource, respond, reuse. London-based talent collective Re Design is a socially aware, not-for-profit design organization that realizes these areas of change.

Re Design creates change not just from designing new products, but by celebrating redesign with a refreshing movement away from the more obvious eco products. It brings together a collective of people—from established and award-winning eco-design talents to up-and-coming designers—for an annual show held during London Design Week, each September. A collective is often stronger than an individual: the larger voice of a cooperative means that its mutual, underlying theme can be more easily heard.

Re Design formulated a unique classification key that highlights nine different approaches to the design of a diverse collection of products. During the show, the visitor can follow colored codes and relevant commentary to identify the concepts behind each product. For example, Ure Chair is classified as Recycle and Reduce; Breathe Bench and Planter as Reclaim and Resource; How Green Is My Valet? as Resource; Me Old China as Reuse, Remind, and Respond; Margo as Reclaim, Recreate, Remind, and Resource; Rockit! as Resource, Respond, and Recreate; Destination Rollerblinds as Remind and Reuse; and Beanpod as Reclaim, Respond, and Reuse. The use of this color key enables the visitor to interact with the show, the

designers, and the bigger question—their purchasing impact on these issues. A well-executed show can make a real difference and create much more than just a platform to view the latest design. This show makes us question the meaning of "good" design.

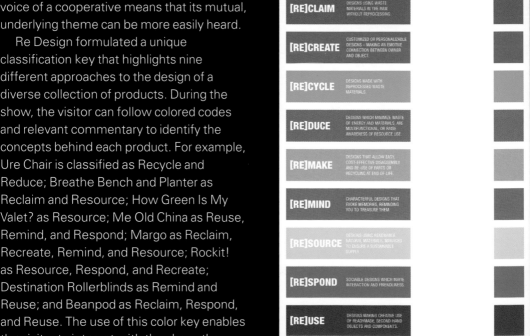

[RE]CLAIM — DESIGNS USING WASTE MATERIALS IN THE RAW WITHOUT REPROCESSING.

[RE]CREATE — CUSTOMIZED OR PERSONALIZABLE DESIGNS – MAKING AN EMOTIVE CONNECTION BETWEEN OWNER AND OBJECT.

[RE]CYCLE — DESIGNS MADE WITH REPROCESSED WASTE MATERIALS.

[RE]DUCE — DESIGNS WHICH MINIMIZE WASTE OF ENERGY AND MATERIALS, ARE MULTIFUNCTIONAL, OR RAISE AWARENESS OF RESOURCE USE.

[RE]MAKE — DESIGNS THAT ALLOW EASY, COST-EFFECTIVE DISASSEMBLY AND RE-USE OF PARTS OR RECYCLING AT END OF LIFE.

[RE]MIND — CHARACTERFUL DESIGNS THAT EVOKE MEMORIES, REMINDING YOU TO TREASURE THEM.

[RE]SOURCE — DESIGNS USING RENEWABLE NATURAL MATERIALS, MANAGED TO ENSURE A SUSTAINABLE SUPPLY.

[RE]SPOND — SOCIABLE DESIGNS WHICH INVITE INTERACTION AND FRIENDLINESS.

[RE]USE — DESIGNS MAKING CREATIVE USE OF READYMADE, SECOND-HAND OBJECTS AND COMPONENTS.

Right: Ure chair
Designed by Cohda. Plastic
waste from domestic use
is formed into HDPE (high
density polyethylene) as a
"new" raw material. This is
heated and reformed into
a unique, sculpted chair.

**Below: Breathe Bench
and Planter**
Designed by Greenpiece. An
integrated planter and hidden
catchment system nourishes a
plant with rainwater collected
from below the bench seat.
The bench is made from FSC-
certified timber.

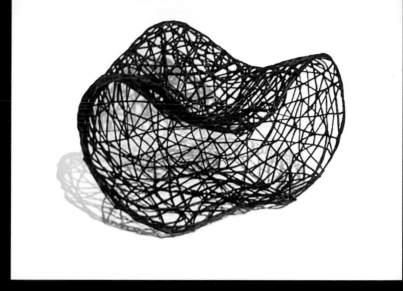

**Left: Re Design
classification key**
This color classification key
prompts users and designers
to think about what constitutes
good design.

How Green is my Valet?
Designed by David Henshall
Design. Shoe and jacket tidy
made from sustainable, FSC-
certified beech. Plant oil finish.

Me Old China
Designed by People Will
Always Need Plates. Odd
pieces of lost or unwanted
dinner services find a new
life within this cake stand.

Margo

Designed by Beef Design.
Matting made from hemp, one
of the truly universally adaptable
materials, is molded together
with a stylish, cast-off fabric
to create a coffee table with
an integrated magazine rack.

TENHAM COURT ROAD | ST PAUL'S CATHE

DDINGTON GREEN
CE OF WALES HARROW ROAD
RUBS LANE HARROW ROAD
ONEBRIDGE PARK
SUDBURY SWAN
BURY TOWN STN
WEMBLEY EALING ROAD

Roehampton
Kingston Vale 85
Putney High Street
Upper Richmond Rd 264
Dover House Road

Fleet St. Piccadilly
South Kensington Fulham N14
Dawes Road Putney
PRIVATE
To hire a bus or coach
ring 01-222 0033

PICCADILLY C
QUEENSTOWN ROAD BATTERSEA
ROSEBERY AVEN
SLOANE SQ
STREATHAM HILL
STREATHAM H
TOTTENHAM COURT R

Above: Destination Rollerblinds
Bus destination blinds are given a new life, with the "same" function.

Left: Rockit!
Designed by WEmake. DIY rockers for a garden bench. Follow the template to create your very own relaxing additions to a bench. Ready-made rockers are also available.

Right: Beanpod
Designed by David Henshall Design. A bean sprouter made with a re-used bottle base.

Marcel Wanders and Wanders Wonders

Marcel Wanders is more than a designer. He is an extraordinary character with a talent for creating the unexpected. He has become one of the most influential design gurus of the 2000s, working within many creative spheres including product and interior design, architecture, corporate identity, and art direction.

Working as art director for Moooi (see page 166) and as a freelance designer for some of the world's best-known brands, on top of heading up design consultancy Wanders Wonders, he expands the design horizon. Not only is Wanders an exceptional designer, he is also a talented marketeer and a shrewd businessman.

Graduating from the Academy of Arts in Arnhem in 1988, Wanders freelanced until the 1990s when he became one of the founders of über design collective, Droog Design. Droog, meaning "dry" in Dutch, is a collective that nurtures creativity and creatives. This philosophy of nurturing and disseminating knowledge is intrinsic to Wanders' personal approach, evidenced in the mentor strategy of all the businesses he is associated with.

The range of Wanders product portfolio is extensive and reveals something of his design process. Some of his creations include storybooks, sculptures, interiors, accessories, furniture and lighting, wallpapers, tiles, linoleum, and audio electronics. All of his products reveal the story of their own births. Designers are always asked about their inspiration and the origination of their concepts, but with Wanders' collection, the answers are inherent in the pieces themselves.

Many designers reference technology and materials as key influencers of form; others nature or an observation of people's lives and the way they live; and still others intuition, research, or moments of epiphany. Within the Wanders portfolio, all of the above apply in varying degrees. Wanders himself often comments on a dream or a story as a starting point. "Every design has to grow, to be its own individual and create its own rules." Each piece does seem to have "grown" in a different way from the others—each is individual, yet with a definite signature style.

Despite his design eminence, Wanders considers himself an amateur. He approaches each design with fresh, clear eyes to find something that may otherwise have been missed. Each piece is so deeply personal, so revealing of Wanders' internal processes that the user feels connected to the designer.

**Pandora side table,
Merlin antenna, and
Egg satellite speaker**
Designed for HE (Holland
Electro). These pieces form
part of a collection of
multimedia home products
that synthesize technology
and decorative detail. The
Pandora side table includes
an integrated woofer.

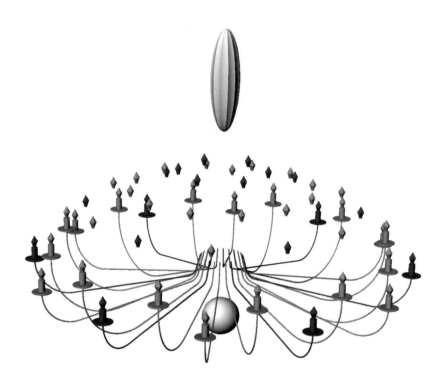

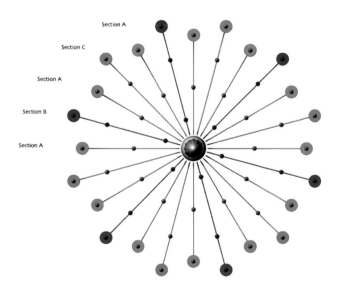

Section A
Section C
Section A
Section B
Section A

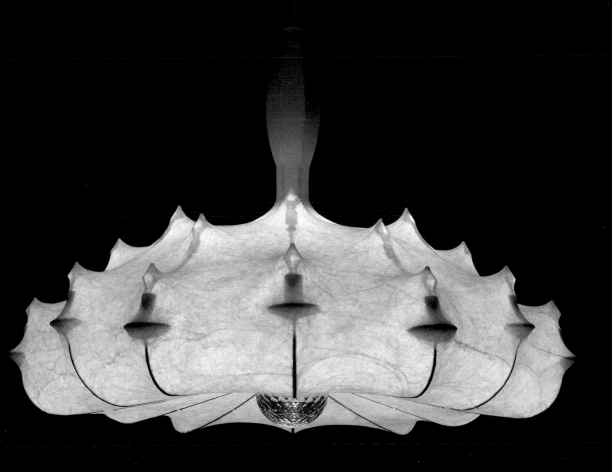

MORE LONG AND THIN

LESS DIAMETER

Zeppelin Suspension Light

Designed for Flo. This design pays homage to one of the best-known and most talented lighting designers, Achille Castiglioni. Wanders shows sensitivity to the original Cocoon collection from the 1960s, but also creates one of the most dramatic pendant lights to date. A membrane diffuses the light and disguises the light source while a chandelier with traditional references gives structural form to this skeletal creation, finished with a large lead crystal ball. The process of design can be seen with the penned amendments in black marker. 3-D software enables the designer to view all angles to see where amendments are needed to create harmony and balance in the structure.

Silver Egg vase
Designed for Droog Design. This chromed porcelain vase, molded by stuffing boiled eggs into a latex condom, was produced in a limited-edition.

Knotted chair
In 1996, this chair caused a
stir for its innovative marriage
of traditional macramé
techniques and technology.
The lightweight chair is hand-
crafted with rope, aramide,
and carbon, then impregnated
with epoxy and hung in a frame
to harden. Gravity gives it
shape and individuality.

Resources

Resources for inspiration and research extend from printed matter, such as books and magazines, to Web-based information in the form of specialist resource libraries, e-zines, company Web sites, and blog diaries.

Here I have listed a handful of books, magazines, and Web sites for research purposes.

Books

Design Secrets: Products. Edited by Industrial Designers Society of America. Rockport, 2001.

Design for Sustainability: A Source Book of Integrated Eco-logical Solutions. Birkeland, Janis. Earthscan Publications Ltd, 2002.

Light Years Ahead: The Story of the PH Lamp. Edited by Tina Jørstian and Poul Erik Munk Nielsen. Louis Poulsen, 1994.

Massive Change: A Manifesto for the Future of Global Design. Mau, Bruce. Phaidon Press, 2004.

Skin: Surface, Substance and Design. Lupton, Ellen. Laurence King Publishing, 2002.

Spoon: 100 Designers, 10 Curators, 10 Design Classics. Edited by Ron Arad, Giulio Cappellini, Ultan Guilfoyle, Brooke Hodge, Laura Housley, Hansjerg Maier-Aichen, Ryu Niimi, Ramon Ubeda, Stefan Ytterborn, Lisa White. Phaidon Press, 2002.

Magazines

Arcade
www.arcadejournal.com

Arena

Blueprint
www.blueprintmagazine.co.uk

Butter
www.mmmbutter.it

Case Da Abitare
www.abitare.it

Cassa

Creative Review
www.creativereview.co.uk

Domestica

Domus
www.domusweb.it

Dwell
www.dwellmag.com

Elle Decor
www.elledecor.com

Elle Wonen
www.ellewonen.nl

Frame
www.framemag.com

FX
www.fxmagazine.co.uk

Icon
www.icon-magazine.co.uk

Idee

Interiors

Interni
www.internimagazine.it

Living Etc
www.livingetc.co.uk

Maison Francaise

Metropolis
www.metropolismag.com

Neo2
www.neo2.es

Objekt
www.objekt.nl

Ottagono
www.ottagono.com

Sense

Spoon
www.spoonlive.com

Surface
www.surfacemag.com

Vogue
www.vogue.com

Wallpaper*
www.wallpaper.com

Competitions

Braun Prize
www.braunprize.com

Designboom (various competitions)
www.designboom.com

Good Design Award
www.g-mark.org

IF Concept Award
www.ifdesign.de

Materialica Design Award
www.materialicadesign.de

Muji Award
www.muji.net/award

Promosedia International
Design Competition
www.promosedia.it

Red Dot Design Award
www.red-dot.de

Web sites

Architonic

www.architonic.com
The independent source for products, materials, and concepts in architecture and design.

Core77

www.core77.com
Includes articles, discussion forums, an extensive event calendar, portfolios, job listings, and a database of design firms, schools, vendors, and services.

Designboom

www.designboom.com
E-zine featuring articles on art, architecture, fashion, photography, and graphics.

Design-engine

www.design-engine.com
Multipurpose portal with interviews and links to industrial designers, furniture makers, and showrooms.

Design*Sponge

www.designsponge.blogspot.com
Daily Web site dedicated to home and product design. It features store and product reviews, sale and contest announcements, new designer profiles, trend forecasting, store and studio tours, student design, and international design shows.

Dexigner

www.dexigner.com
Includes design news, competitions, forums, and links.

Places and Spaces

www.placesandspaces.com
Online store selling designer furniture and accessories.

MoCo Loco

www.mocoloco.com
Web magazine featuring contemporary design news and views.

Stylepark

www.stylepark.com
Extensive international product archive for design and architecture.

Glossary

anthropometrics
This is the study of human measurement. It involves taking accurate measurements of the size and proportions of the human body, as well as factors such as reach and visual range. Anthropometrics gives designers the information they need to properly size products to "fit" the end user.

Art Nouveau
This design movement began in Europe in the late 1880s. The followers of art noveau rejected historicism in favor of new forms, and embraced mass-production. Art Nouveau designs were commonly inspired by nature and feature organic, flowing shapes. The style had many regional variants, including Jugendstil in Germany.

batch production
Batch production refers to the production of a set number of a particular product rather than a one-off item, or the continuous production of a particular item. It can reduce initial capital outlay as it allows one set of equipment to be used for the production of more than just a single product. It also allows for the trial run of a product. However, it does require careful planning and scheduling to avoid costly changes (which are difficult to make once a batch is in production), build-up of works in progress, and downtime.

Bauhaus
The Bauhaus design school was founded in 1919 by architect Walter Gropius, in Weimar, Germany. The school's manifesto was to establish architecture as the dominant design forum, to elevate the status of craft skills to that of the fiine arts, and to improve industrial products through the combined efforts of artists, industrialists, and craftsmen. Bauhaus designers favored functionality over ornamentation; they believed form should follow function.

brief
This identifies predetermined guidelines and sets the agenda for the design process of a project within the context of production, marketing, and accounting. The objective of the brief is both to inspire the designer and to assist in the realization of the final product.

computer-aided design (CAD)
A design process that utilizes a computer-graphics program for drafting and visualization in 2-D or 3-D space. CAD is primarily used for the creation of graphics, but also allows information regarding costs and materials to be input and stored.

core competencies
The core competencies of a design company are the specific areas of knowledge and skills it has that gives it a competitive advantage in the marketplace. These are studio-wide capabilities based on the combined skills of all individuals and teams involved in the design process.

customization
In terms of product design as explored in this book, customization refers to the altering of a product through a design innovation, subtle or major, to give it a unique character, a perceived added value, or a new application.

Delftware
A style of glazed earthenware, usually white with blue decoration, named after the town of Delft in the Netherlands, which was the original production area.

end user
The end user is the final consumer of a finished product; the ultimate target market. Isolated from the design and production process, the rationale for the end user's satisfaction is that the product performs the tasks for which it was purchased.

ergonomics
This is an applied science concerned with designing and arranging products so that the end user and the product interact safely and in the most efficient way.

flatpacks
Products, mostly furniture, that are delivered to the retail outlet and/or end user in a "flat," compact package for subsequent assembly. Flatpacks allow ease of delivery and result in cost-savings by reducing the volume of the packages to be transported by the distributor and retailer. They also reduce the storage space required.

functionality
Products are described in terms of the specific tasks they have been designed to perform; the functionality of a product refers to its success in meeting the consumer's needs in performing these tasks. Consumer demands are related not only to the purely functional, but also to the esthetic and experiential benefits of a product.

globalization
The increasing cross-cultural influence and interaction between cultures and consumers around the world. Product designers are now in a position to make their products international in appeal and application.

Golden Mean
Also known as the Golden Section, the Golden Ratio, the Divine Proportion, the Golden Rectangle, and the Fibonacci Sequence. The Golden Mean is a tool for achieving pleasing proportion in design. In relation to a line divided into two segments or the length and width of a rectangle and their sum, the ratio of the whole to the larger part is the same as the ratio of the larger part to the smaller.

hybrid product
A hybrid product is formed when two or more objects are merged to create something new. This new item will still have visible references to the original products.

industrial design
The field of developing physical design solutions to meet a particular need, and/or to improve the esthetics and usability of products. Aspects for consideration include the shape of the object, location of details, proportions, colors, textures, sounds, and the product ergonomics.

information flow
Information flow refers not only to the effective communication and sharing of information between individuals involved in the design process, but also to the effective research into, and access to information on the end user's requirements. A product designer should be provided with as much information as possible about their customers' product requirements in order to create the best design solution.

injection molding
This process involves the feeding of polymer pellets into a machine through a hopper and then into a heated barrel. The heat from the barrel turns the plastic into a liquid resin, which is then injected into the mold. This allows designers the freedom to create almost any imaginable form.

intellectual property
This refers to systems of legal protection for particular means of expressing information and ideas. Examples include patents, copyrights, industrial design rights, and trademarks.

Jugendstil
The German interpretation of Art Nouveau, prevalent during the 1890s. Jugendstil designers advocated the use of natural forms.

life cycle
The various stages through which a product passes from introduction to the market to withdrawal or obsolescence. There are many considerations that relate to a product's life cycle. These include: growth and decline of market, changes in style and fashions, availability, reliability, systems for maintenance, opportunities for upgrades, and ecological considerations.

logistics
Management of the details of production and distribution of a product. Logistics encompasses many supply issues: raw material supply and labor accessibility, the flow of materials or finished components within the production process, and the questions associated with timely fulfilment to a specific market. Logistics are often not incorporated into the design process with the result that, while product design can contribute to efficient global logistics, it more often causes major limitations.

Memphis period
Memphis was the name of the Milan-based collective of product and furniture designers led by Ettore Sottsass. Its work was characterized by brightly colored plastic laminates with kitsch geometric, and leopard-skin patterns.

minimalism
A design movement first described in the mid-1960s, the key characteristics of which include pared-down forms, use of industrial materials, geometric forms, grid-based compositions, and rigid planes of color.

modeling
A tool within the process of concept visualization that offers either a hands-on or a computerized methodology.

Modern Style
A clean, streamlined design style from the 1930s characterized by strong geometric shapes, asymmetry, and polished surfaces.

Modernism
The leading design movement of the 20th century. Key characteristics include simple, undecorated forms; smooth finishes; minimal surface modeling; and use of state-of-the-art materials.

multifunctionality
The aspect of products that offer more than one form and/or use. Additional functionality can add value to products by creating perceived improvements in time- and space-saving properties.

Munsell system
Developed by American artist A. H. Munsell, this system was designed to classify standards for printing inks, color pigments, and artists' paints. It uses a collection of color charts made up of printed color swatches for each hue. Each color is assigned a number to define the hue (which determines the chart on which it appears). Further values for luminance and saturation give vertical and horizontal coordinates to indicate where it appears on this chart. By using these three values, you can find the color on the Munsell chart and use it to ensure that it remains constant throughout an imaging system.

obsolescence
The process or cycle by which products cease to be useful or fall out of style. Obsolescence may be due to changing fashions or technological advances. Many products have a planned obsolescence incorporated into their design in order to induce consumers to buy a new model of the product and thus create a continuous demand.

packaging
The covering or container in which a product is transported and displayed. A product's packaging should be considered in terms of logistics (ease and cost of distribution and storage), protection (safe and clean delivery of the product to the consumer), and marketing (it should heighten the product's appeal to the public).

patent
A form of legal protection/ownership for products or processes featuring new functional or technical aspects. Patents are concerned with what products do and how they do it, what they are made of, or how they are made. A patent holder has an exclusive right to apply a particular invention or even to prevent any other party from using it. The usual term is around 20 years.

Postmodernism
A stylistic response to Modernism. Key characteristics are: a rejection of industrial process; a merging of past styles; surface decoration; layered imagery, collage, and photomontage.

product design
A subfield of industrial design. Product design utilizes a number of varied processes to develop physical solutions to specific needs. Products may be high-tech or low-tech. They may be produced in an exclusive limited-edition, batch produced, mass-produced, custom-made, or customized.

product designer

Product and industrial designers design all the things people use in their daily lives, from clocks to chairs, CD players to vases, vacuum cleaners to cars. Their role is to consider how to make items easier to use, more efficient, cheaper to produce, or more esthetically pleasing. They consider all of the experiential benefits of the product. This requires a careful exploration of what is required, and investigation into what the end users' needs and demands are. Product designers research and develop ideas in order to come up with a design solution for a new product, or to improve the function or visual appeal of an existing product. The process involves: taking a brief, making initial sketches, preparing detailed drawings, making samples or working models, and testing the design. Every stage involves working with other skilled professionals including trend forecasters, engineers, model-makers, logistics experts, and marketing managers. Product designers require a knowledge of technology, production methods, materials, and market trends. In addition to this hands-on design work, product designers are also involved in meetings with clients and presenting their ideas. They may also be involved with putting together bids and proposals for work.

prototype

An early form for a new design or product. The prototype serves as the basis for, and evidence of need for, any later developments and changes.

rapid prototyping

Rapid prototyping is used for the modeling of components that require absolute precision engineering.

rotational molding

A process whereby a hollow mold is filled with a powder resin, then rotated biaxially in an oven until the resin coats the inside of the mold and cures. The mold is then cooled and the product or part removed. Rotational molding is usually used to create hollow, large-scale products. It has relatively low tooling costs making it ideal for low production-run pieces.

reappropriation

The use and application of materials or found items in a way other than their original intended purpose to create new products and/or meanings.

recycling

Reforming or adapting materials in order to make them available for use in a new product. In product design there are two main considerations regarding recycling: whether it is recyclable (can the product be recycled once it is no longer of any use) and whether it is formed from recycled or reappropriated materials.

Index

Credits

Firstly thanks to everyone who I might forget to mention!

Thanks to all those who kindly sent images and answered questions for *What is Product Design?*, especially those companies and designers who gave us a little more than usual, with sketches and photographs that illustrate design in progress.

Thanks to Lindy Dunlop for her wonderfully calming and efficient coordination, from start to finish, in the writing of this book.

Thanks to Mark Cox and Susanna Moore for their endless image-collecting duties which never seemed to end.

Thanks to the photographers, graphic designers, image manipulators, production coordinators, and set builders etc. who have supported Places and Spaces in ongoing projects: Ian Rippington, John Reynolds, Sophie Broadbridge, Keith Stephenson, Andrew Allen, Mike Teesdale, Al Roots.

Most of all a massive thank-you for all your support to Shaggy (my husband), mum and dad, sister (Clare), and amazing friend (Sarah). Often my ability to think clearly and put pen to paper failed and it was hard to keep going when the end seemed so far away, so thanks for everything.